CUBICAL QUAD ANTENNAS

THIRD EDITION

WILLIAM ORR, W6SAI and **STUART COWAN,** W2LX

RADIO AMATEUR CALLBOOK
P.O. BOX 2013 LAKEWOOD, NEW JERSEY 08701

Publisher's note: *Cubical Quad Antennas,* Third Edition, was previously published as *All About Cubical Quad Antennas,* New Third Edition.

Caution: Working on antennas or towers can be dangerous. All warnings on the equipment and all operating and use instructions should be adhered to. Make sure that the antenna is disconnected from the station equipment before you begin to work on it. Make sure that your antenna is not close to power lines, and that it cannot drop on a power line if wires or supports fail. Do not attempt to climb a tower without a safety belt. It is best to work on your antenna with someone who can assist you with tools and will be able to help in the event of a problem or an emergency.

Published in 1993 by Radio Amateur Callbook
(an imprint of Watson-Guptill Publications,
a division of BPI Communications, Inc.),
P.O. Box 2013, Lakewood, New Jersey 08701

Library of Congress Catalog Card Number: 82-80282
ISBN 0-8230-8703-4

Manufactured in the United States of America

2 3 4 5 6 7 8 9/01 00 99 98 97 96 95 94

TABLE OF CONTENTS

FOREWORD

Born high in the South American Andes many years ago, surrounded by a civilization thousands of years old, the Cubical Quad antenna has caught the interest of amateurs the world over and has taken its place beside the more sophisticated antenna arrays conceived in the electronic laboratories of the world.

The concept of the famous "Quad" was a stroke of original thought, conceived by an amateur in one of the little known areas of the world in an attempt to solve a problem that could not be solved! The success of the Cubical Quad—the brainchild of W9LZX—in overcoming the myriad difficulties of antenna design for a tropical shortwave broadcasting station is worthy inspiration for any amateur. The story of the Quad spells out the true ham radio spirit of "make do" when confronted with a problem that cannot be solved.

This, then, is the story of the Quad antenna; its humble beginning, what it does, how it works, and its spectacular success in the world of amateur radio.

The third edition of this Handbook provides updated information on the Quad antenna derived through additional on-the-air tests and recent field strength measurements made on model test ranges. Gain of the Quad loop has been reevaluated in light of these tests and new gain figures derived for various Quad configurations, confirming independent measurements made by other experimenters. Grateful thanks are extended to those many amateurs for their help and assistance in the preparation of this Handbook.

CHAPTER I

The Story of the Cubical Quad Antenna

The Cubical Quad is an unusual antenna, and it has a unique and interesting history. The development and growth of the ordinary amateur antenna follows a rather stereotyped story. The theory of the antenna usually makes its first bow in some technical publication, such as the *Proceedings of the I.E.E.E.* Next, the antenna is used and tested by some radio engineer who is also an ardent amateur. Soon, by word-of-mouth, the story of the antenna spreads and eventually it is publicized in some amateur journal. During the growth and development of the antenna, the story is embellished with tales of fantastic gain, unbelievable front-to-back ratio, and other magical attributes possessed by this antenna which no other antenna can lay claim to. Over a period of years after the hue and cry has dimmed a bit, the antenna either falls into limbo and is forgotten, or it takes its rightful place in the great group of popular amateur equipment. Meanwhile, some other new development has probably surpassed the antenna in the interest of the amateur.

An exception to this story is the Cubical Quad antenna. Springing full-grown, as it were, into popularity with no formal engineering ancestry, the Quad has been simultaneously hailed as the greatest antenna development of the age, and damned as the greatest hoax of the century. Naturally, the truth lies somewhere between these two violent extremes.

In order to arrive at an unbiased opinion of the antenna, it is necessary to examine its past history, determine the method of operation, and arrive at a proper method of feeding the array.

EARLY HISTORY OF THE QUAD

In the year 1939 a group of radio engineers from the United States traveled to the South American republic of Ecuador to install and maintain the Missionary Radio Station HCJB, at Quito, high in the Andes mountains. Designed to operate in the 25 meter shortwave broadcast band with a carrier power of 10,000 modulated watts, the mission of HCJB was to transmit the Gospel to the Northern Hemisphere, and to tell of the missionary work in the wilds of Ecuador. To insure the best possible reception of HCJB in the United States a gigantic four element parasitic beam was designed, built, and erected with great effort and centered upon the heartland of North America.

The enthusiasm of the engineers that greeted the first transmission of Radio HCJB was dampened after a few days of operation of the station when it became apparent that the four element beam was slowly being destroyed by an unusual combination of circumstances that were not under the control of the worried staff of the station. It was true that the big beam imparted a real "punch" to the signal of HCJB and that listener reports in the path of the beam were high in praise of the signal from Quito. This result had been expected. Totally unexpected, however, was the effect of operating the high-Q beam antenna in the thin evening air of Quito. Situated at 10,000 feet altitude in the Andes, the beam antenna reacted in a strange way to the mountain atmosphere. Gigantic corona discharges sprang full-blown from the tips of the driven element and directors, standing out in mid-air and burning with a wicked hiss and crackle. The heavy industrial aluminum tubing used for the elements of the doomed beam glowed with the heat of the arc and turned incandescent at the tips. Large molten chunks of aluminum dropped to the ground as the inexorable fire slowly consumed the antenna.

The corona discharges were so loud and so intense that they could be seen and heard singing and burning a quarter-mile away from the station. The music and programs of HCJB could be clearly heard through the quiet night air of the city as the r-f energy gave fuel to the crowns of fire clinging to the tips of the antenna elements. The joyful tones of studio music were transformed into a dirge of doom for the station unless an immediate solution to the problem could be found.

It fell to the lot of Clarence C. Moore, W9LZX, one of the engineers of HCJB to tackle this problem. It was obvious to him that the easily ionized air at the two mile elevation of Quito could not withstand the high voltage potentials developed at the tips of the beam elements. The awe-inspiring (to

The studios of HCJB located in Quito, Ecuador, the birthplace of the Cubical Quad antenna. The simple Quad was used for many years when the transmitters of HCJB were located in the city. A few years ago the transmitters were moved to a 45 acre site at Pifo about 15 miles east of Quito. Programs originating in the studio are sent via frequency modulation link to the transmitting site. The old Quad antenna has been replaced with a steel tower that supports the FM antenna.

the natives) corona discharges would probably disappear if it were possible to operate HCJB at a sea level location. This, however, was impossible. The die was cast, and HCJB was permanently settled in Quito.

What to do? Moore attacked the problem with his usual energy. He achieved a partial solution by placing six-inch diameter copper balls obtained from sewage flush tanks on the tips of each element. An immediate reduction in corona trouble was noted, but the copper orbs detuned the beam, and still permitted a nasty corona to spring forth on the element tips in damp weather. Clearly the solution to the problem lay in some new, different approach to the antenna installation. The whole future of HCJB and the Evangelistic effort seemed to hinge upon the solution of the antenna problem. The station could not be moved, and the use of a high-gain beam antenna to battle the interference in the crowded 25 meter international shortwave broadcast band was mandatory. It was distressingly apparent to Moore that the crux of the matter was at hand.

THE BIRTH OF THE QUAD

In the words of W9LZX, the idea of the Quad antenna slowly unfolded to him, almost as a Divine inspiration. "We took about one hundred pounds of engineering reference books with us on our short vacation to Posoraja, Ecuador during the summer of 1942, determined that with the help of God we could solve our problem. There on the floor of our bamboo cottage we spread open all the reference books we had brought with us and worked for hours on basic antenna design. Our prayers must have been answered, for gradually as we worked the vision of a quad-shaped antenna gradually grew from the idea of a pulled-open folded dipole. We returned to Quito, afire with the new concept of a loop antenna having no ends to the elements, and combining relatively high transmitting impedance and high gain."

A Quad antenna with reflector was hastily built and erected at HCJB in the place of the charred four element beam. Warily, the crew of tired builders watched the new antenna through the long operating hours of the station. The vigil continued during the evening hours as the jungle exhaled its moisture collected during the hot daylight hours. The tension of the onlookers grew as a film of dew collected on the antenna wires and structure, but not once did the new Quad antenna flash over or break into a deadly corona flame, even with the full modulated power of the Missionary station applied to the wires. The problem of corona discharge seemed to be solved for all time.

The new Quad antenna distinguished itself in a short time with the listeners of HCJB. Reports flooded the station, attesting to the efficiency of the simple antenna and the strength of the signal. In his spare time, Moore built a second Quad antenna, this one to be used in the 20 meter band at his ham station, HC1JB, in Quito.

*   *   *   *   *

At a later date, after Moore had returned to the United States, he applied for a patent covering the new antenna. The fact that the Quad-type antenna radiated perpendicular to the plane of the loop was deemed by the Patent Office to be of sufficient importance to permit the issuance of a patent to Clarence C. Moore covering the so-called Cubical Quad antenna.

Other shortwave broadcasting stations in the Central American area soon heard of this new, high gain, corona-proof antenna, and Moore built several Quads on order, including a large rotating giant for 49 meter shortwave broadcast work at station TGNA in Guatemala City, Guatemala. This antenna has been used for years with success at an altitude of 5,000 feet.

The outstanding signal of HC1JB in the 20 meter amateur· band quickly flooded Moore with inquiries about his new antenna. Soon, Quad antennas

The new transmitters of HCJB are located at Pifo, Ecuador. The arrays range from six to twenty-four elements, depending upon the desired direction of transmission. Herb Jacobsen, ex-transmitter engineer of HCJB pays tribute to the Quad, saying, "We found the Quad a very useful antenna for locations with limited space and it served us well for many years."

were being used by amateurs on both the 10 and 20 meter band, and the amazing success story of the Quad came into being.

How successful is the Quad antenna? In 1948 W9LZX said:

"Well, we love the little antenna! In addition to solving the corona problem at HCJB and other tropical broadcast stations, the antenna has other commendable attributes. It is very quiet for reception, and gives little trouble from rain static which often plagues three element parasitic beams. The Quad fits into a smaller space than the conventional three element beam, and exhibits a power gain that surely is comparable to a parasitic array of equal or greater size. In addition, the Quad can be matched to a coaxial transmission line, or it may be directly fed with an open wire line. Finally, it is extremely inexpensive to build and simple to assemble. You know, it is not easy to obtain aluminum tubing in many parts of the United States, and practically impossible in some areas of the world. For the amateur with little money or no source of supply of aluminum tubing for a parasitic array, I can't see how he can beat the Quad antenna. It's a honey! We've used it for over a decade, and we know!"

* * * * *

That is the story of the birth of the Quad antenna and its spectacular rise from the wilds of Ecuador to use in radio stations throughout the world. The fame of this unusual antenna has literally spread by word of mouth until it is "topic number one" wherever amateurs discuss antennas and DX.

CHAPTER II

The Quad: How Does It Work?

Many hard questions remain to be answered concerning the Quad antenna, as countless amateurs view it with a degree of skepticism. *How* does the Quad work? *Why* does it work? Is it better than a three element beam—its chief rival? Does it *really* provide ten or twelve decibels of gain as some of its enthusiastic boosters claim? How does one go about building a Quad? These and many other questions will be answered in the following chapters of this Handbook.

PARASITIC AND DRIVEN ARRAYS

The great majority of beam antennas belong to one of two families. That is, they are either described as *parasitic arrays* or *driven arrays*. The difference between the two groups is the method by which the directive elements are excited. A parasitic array is one wherein the directive elements are *inductively coupled* to the radiator. The popular "three element beam" is an array of this type. A driven array is one that has all the elements *directly excited* by the r-f source. A "Lazy-H," or the well-known "W8JK" beam are examples of driven arrays. In addition, an *array of arrays* may be built combining the features of these two families in which some elements are driven directly, and others are parasitically excited. The Cubical Quad antenna falls into this latter category as it employs elements of both types.

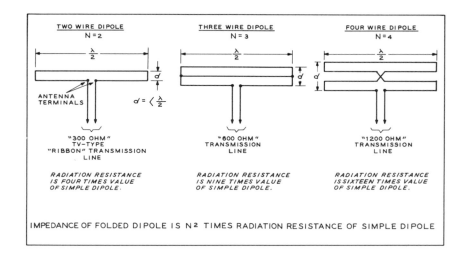

Fig. 1 The ancestor of the Quad loop is the simple folded dipole. This antenna has the same field pattern as the single wire dipole, but presents a much higher value of radiation resistance to the transmission line. The actual value of radiation resistance is a function of the square of the number of conductors in the dipole, multiplied by the radiation resistance of the single wire dipole. Thus, in the case of the two wire dipole, the radiation resistance is four times 72 ohms, or 208 ohms. This antenna may be directly fed with a non-resonant line made of "300 ohm" TV-type ribbon.

The Driven Element of the Quad

For purposes of this discussion, let us examine the driven element of the Quad, forgetting about the parasitic reflector for a moment. It is convenient to borrow the description of the Quad element given by W9LZX — "a pulled-open folded dipole." This is a good starting point for investigation. A simple folded dipole is shown in figure 1. This antenna consists of two or more closely spaced half-wave dipoles connected in parallel at their extremities. One of the dipoles is broken at the center to permit attachment of a balanced transmission line.

The radiation resistance at the center of a single dipole is approximately 72 ohms at the frequency of resonance when the dipole is placed one-half wavelength above a conducting surface. As additional dipoles are brought in close proximity with the original one and are connected in parallel at the extremities, the radiation resistance at the center of the split dipole will rise sharply. A two wire folded dipole has a radiation resistance of four times the value of a single element, or about 288 ohms. A three wire dipole

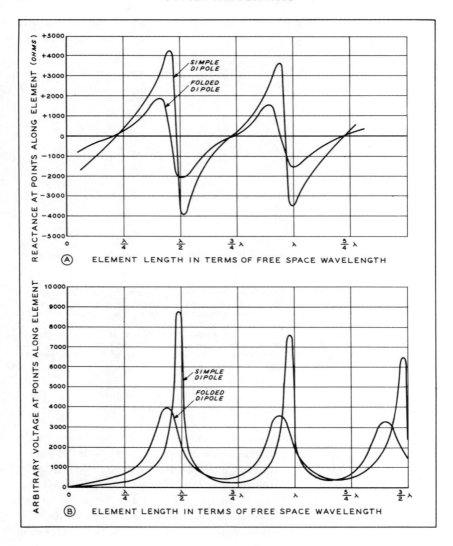

Fig. 2 A plot of the reactance and r-f voltage at various points on an antenna. The points where the reactance curves of figure A cross the zero axis indicate the resonant lengths of the antenna. Because of the end-effects and capacity to ground the half-wave antenna is shorter than the free-space half-wavelength. This foreshortening is increased in the case of the folded dipole as the effective diameter of the antenna is greater. In addition the average impedance values of the antenna are lower, as shown by the curve. R-f voltage at the tips of the element is proportional to the impedance at this point, the folded dipole having a lower voltage for a given power, as compared to the single wire dipole.

has a radiation resistance figure close to nine times the impedance of a single wire dipole, or about 648 ohms. It can therefore be said that the radiation resistance of a multi-conductor dipole is N^2 times the radiation resistance of a single dipole, where N is the number of separate dipoles.

It is interesting to note that the radiation pattern of the folded dipole agrees in all respects with that of the single wire dipole. The angle of radiation and radiated power for the two antennas are equal, providing external forces and fields are equal. An interesting side effect is apparent when the multi-wire dipole is compared to the single wire version. The bandwidth of the single dipole is quite narrow compared to the bandwidth of the folded dipole. That is, the circuit "Q" of the folded dipole is low in comparison to the "Q" of the single dipole. The folded dipole may be thought of as a broad band, low "Q" system as compared to the characteristics of the normal dipole. This means that the impedance at the tips of the folded dipole is much lower than that value noted at the tips of the single wire dipole (figure 2).

This fact is an important consideration when the antenna is used in conjunction with a high powered transmitter. Generally speaking, for a given amount of impressed power, higher values of r-f voltage exist at the high impedance points in any antenna that at the low impedance points, and the amount of r-f voltage at any point is proportional to the antenna impedance at that point. The voltage distribution curves typical of these two types of antennas are shown in figure 2B. The folded dipole has a measurably lower value of r-f voltage at the extremities, and is less susceptible to corona discharge and other undersirable high voltage phenomena.

The "Open" Dipole

The simple folded dipole antenna may be "pulled open" as shown in figure 3A to produce a diamond-shaped loop fed at the bottom point. A four wire dipole may be opened in a like manner, as shown in figure 3B. The radiation patterns of the two diamond configurations will be identical, although the radiation resistance of the double loop will be much higher than that of the single loop. We will therefore confine the discussion to the single turn loop for the time being.

If we continue to stretch the folded dipole past the arrangement of figure 3A the antenna will ultimately become a two wire transmission line one-half wave long, shorted at the far end. The input radiation resistance of the folded dipole is about 288 ohms, and the input resistance of the shorted transmission line is zero ohms. It would seem to be reasonable to

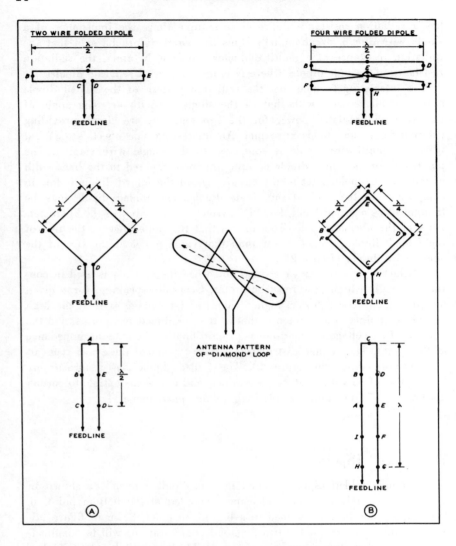

Fig. 3   For purposes of illustration, the two wire folded dipole may be "pulled open" to form a diamond-shaped loop fed at the bottom point. If this distortion of the loop is continued the antenna will become a shorted transmission line. Corresponding points on the dipole are marked in the loop and transmission line cases. The same analogy may be applied to the four wire dipole, which produces a two turn diamond-shape loop when it is elongated. This loop produces the same field pattern as that of the single turn loop, although the radiation resistance is quite a bit higher. The line of maximum field strength of the loop is at right angles to the plane of the loop. When the loop is fed at the bottom the field is horizontally polarized.

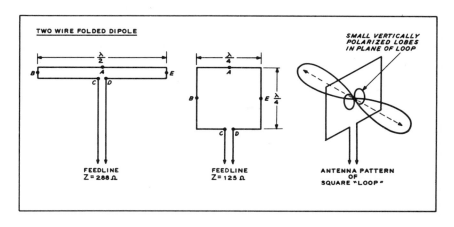

Fig. 4   The folded dipole may be formed into a square loop having two parallel elements, one above the other. The pattern of this loop is horizontally polarized with two small vertically polarized "ears" at right angles to the main lobes. This loop produces a power gain of about 1.4 decibel over the field of the dipole antenna (3.5 decibel gain over isotropic). Loop impedance is about 125 ohms.

assume that the radiation resistance of the "pulled open" dipole to be some intermediate value in the neighborhood of 144 ohms. Measurements made on model antennas indicate this figure is very nearly correct.

This diamond shaped antenna exhibits the radiation pattern of the simple dipole antenna. It has a "figure-8" pattern *and has a power gain of about 1.4 decibels over a dipole* (3.5 decibels gain over isotropic).

It is possible to distort the shape of the folded dipole in another manner, as shown in figure 4. A loop antenna is formed having two parallel elements, one above the other. The square thus formed is fed by a balanced transmission line at the center of the lower element. Each side of the square is approximately one-quarter wavelength long, and the high impedance points of the loop fall at the mid-point of the vertical sides. It is important to note that a loop of this configuration exhibits almost 1.4 decibels power gain, equal to that gain provided by the diamond shaped loop of figure 3A. The radiation pattern of the square, horizontal loop is similar to that of a horizontal dipole except for a slightly narrower lobe perpendicular to the plane of the loop, and slightly reduced radiation in the directions of a line passing through the center of the upper and lower conductors. In addition, a small *vertically polarized* lobe appears at right angles to the main lobe, caused by a small amount of radiation from the vertical wires of the loop. The radiation resistance of this type of loop is approximately 125 ohms, as measured at a height of 0.65 wavelength to the center of the loop.

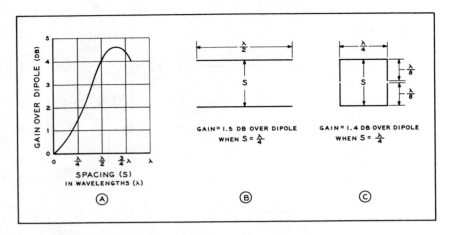

Fig. 5 The power gain of two half-wave elements operating in a vertical stack is shown in graph A. At a separation of one-quarter wavelength, the power gain is 1.5 decibel. A fraction of this gain is lost when the ends of the elements are folded towards each other as shown at C. Since there is no current flowing at the element tips, they may be connected together to form a closed loop. The loop may be square or diamond shaped and is broken at the bottom for feeding.

Power Gain of the Square Loop

The square loop is an interesting antenna, since it also provides about 1.4 decibels power gain — equal to that gain provided by the diamond shape. This useful power gain obviously results from the coupled directivity provided by the upper and lower sections of the antenna which are operating in-phase. The power gain of two half-wave elements operating in phase in a vertical stack, expressed in terms of the vertical spacing is illustrated in figure 5. In this instance, the separation between the upper portion of the loop and the lower portion is 0.25 wavelength. Two half-wave elements spaced this amount would provide a broadside gain of about 1.5 decibel. A small fraction of gain may be lost because the ends of the half-wave elements are bent towards each other to form a closed loop. The single square loop may therefore be thought of as a pair of vertically stacked, horizontally polarized elements spaced 0.25 wavelength, with their extremities connected for feeding purposes. The loop is excited at the center of the lower section at the point of maximum current.

Variations of the Driven Loop

The simple driven loop antenna appears in various forms in the technical literature, although sometimes in clever disguises. A popular European form of driven loop is the so-called "slot antenna" shown in figure 6A. Thought by many to be some form of slot-excited affair, this stacked beam employs two dipoles that are connected at their tips. In order to obtain greater stacking gain the separation between the upper and lower sections is increased at the expense of the radiating section of the two driven elements. Another version of the quarter wavelength loop is the "Bruce antenna," composed of quarter-wave loops connected together, as shown in figure 6B.

The sides of the loop may be increased to one-half wavelength. Many forms of antennas can be constructed from loops of comparable size, and among them are the Lazy-H antenna and the Sterba curtain. All of these exotic antennas, however, are cousins under the skin to the simple square loop of the Cubical Quad antenna.

Adding a Parasitic Element to the Quad Loop

It is possible to add a second Quad element functioning as a parasitically excited reflector or director in front of the Quad loop, as shown in figure 7.

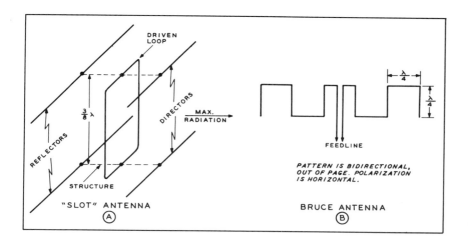

Fig. 6   Variations of the quarter-wave loop are shown above. At **A** is the so-called slot antenna which is basically a Quad loop having single element parasitic reflectors and directors. The tips of the driven element are folded towards each other and connected at their tips. In order to obtain greater stacking gain the separation between the upper and lower sections is increased at the expense of the radiating sections of the driven elements. At **B** is shown the Bruce antenna. This beam is made of sections of quarter-wave loops arranged in line and fed at the center. The vertical sections are out of phase, and the vertical radiation is largely cancelled. Gain of the Bruce beam is low considering its length.

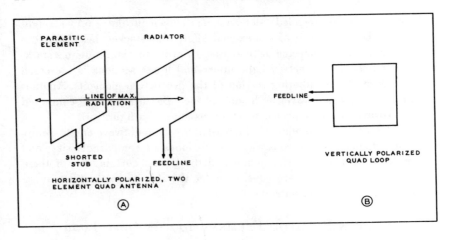

Fig. 7　The Quad antenna is formed from two horizontal loops placed as shown in illustration A. Maximum radiation is in a line perpendicular to the plane of the loops. The loop serving as the parasitic element usually has a shorted stub in place of the feedline. The stub length may be varied to tune the array for maximum gain at the operating frequency. If the driven loop is fed at one side, as shown at B the field of radiation is vertically polarized. This type of loop is often used at 2 and 6 meters. The size of the parasitic element may be suitably altered so that the shorted stub shown above is not required.

When a parasitic element is added to a simple dipole, the achievable power gain is a function of the spacing and tuning of the parasite, and *also* a function of the "Q" or inherent selectivity of the parasite. The effectiveness of the parasitic element is proportional to the coefficient of the coupling between the element and the radiator. In general, the higher the "Q" of the parasitic element, the greater the coupling and the greater the power gain of the array. A low-Q reflector, such as a "billboard"-type screen provides a power gain of about 3 decibels, while a high-Q parasitic reflector, such as a thin wire can provide a power gain close to 6 decibels. In all experimental cases it has been found impossible to secure as high a forward gain figure with low-Q parasitic elements as with high-Q elements. In the case of long parasitic arrays having many directors, this means that antenna gain will be the highest when employing the thinnest possible parasitic elements, until the point is reached where the surface conductivity becomes poor enough to materially affect the Q of the elements.

By its configuration, the square loop has a rather high value of radiation resistance and its Q is somewhat lower than that of an equivalent dipole. Even so, the addition of a parasitic loop to the Quad element raises the overall gain figure by nearly 6 decibels. The maximum power gain of the two element Quad antenna is therefore the sum of the loop gain and the parasitic gain, or $1.4 + 5.9 = 7.3$ decibels, as compared to a dipole. This is approximately equal to the power gain of a "three element" Yagi-type parasitic beam. The actual power gain of a typical Quad antenna may be somewhat lower than the maximum theoretical gain due to r-f losses in the coupling network (if any) and in the wires of the array. Also, gain is dependent somewhat upon the tuning technique. Element spacing also plays a minor part in gain determination, as covered in the next chapter. In any event, the two element Quad turns out to be stiff competition for the three element Yagi. A tri-band Quad, moreover, usually outperforms a tri-band "trapped" Yagi because of the lower r-f losses of the former antenna.

Wave Polarization

When the Quad loop is fed at the bottom the points of maximum current occur in the two horizontal wires. The current flowing in the vertical wires is considered to be out of phase, and the radiated field is correspondingly small. Even so, a small vertically polarized field exists around these wires. If the Quad loop is rotated through 90 degrees in its plane, the feed point will be in the middle of one side, and the wires that formerly were in the vertical plane are now in the horizontal plane. The main field of the loop is now vertically polarized, with a small field of horizontal radiation about the horizontal wires, as shown in figure 7B.

Since the polarization of the radio wave is obscured after reflection from the ionosphere, either a vertically or horizontally polarized Quad may be used with good results on the high frequency bands. The vertically polarized antenna, however, will be more receptive to man-made interference (automobile ignition noise, for example) since such radiations are usually vertically polarized.

The Closed Quad Loop

The total loop length of a typical Quad element may be adjusted so that the tuning stub shown in figure 7A is not required. Stub tuning a Quad element is a good way to make preliminary adjustments, to be sure, but for lowest wind resistance and best current distribution in the element it is a good idea to dispense with the tuning stub and to make the loop slightly larger to compensate for the missing stub. Dimensional data for both methods of construction are given in various tables in this Handbook.

CHAPTER III

Characteristics of the Quad Antenna

Any antenna serves as a coupling device to convert electronic energy supplied by the transmitter to electrostatic and electromagnetic waves which are propagated through space. At the receiving station a similar antenna converts the received energy back to electronic energy which can be detected and demodulated by the receiving equipment.

The overall efficiency and operating parameters of the antenna may be expressed in terms of radiation resistance, directivity, power gain, and effective aperture of the antenna.

ANTENNA TERMINOLOGY

Certain terms and characteristics peculiar to antenna systems in general should be defined, and the problems and hazards of determining antenna operating characteristics should be covered before definite gain figures and dimensions are given for the Quad antenna.

Radiation Resistance

The *radiation resistance* of an antenna may be defined as that value of resistance which, when substituted for the antenna, will dissipate the same amount of power as is radiated by the antenna. The actual value of radiation resistance of any antenna is determined by the configuration and size of the antenna, and the proximity and character of nearby objects. The value of radiation resistance bears no relation to the efficiency or power gain of the antenna, and the fact that one antenna has a different value of radiation resistance than that of another antenna does not necessarily mean that the first antenna is better or more efficient than the second, or vice-versa. It is important to know the magnitude of the radiation resistance in order to match it to the nominal impedance value of the transmission line.

Antenna "Q" and Resonance

If a load is connected to an antenna that is energized by a passing radio wave a certain amount of power may be extracted from the wave and will be dissipated in the load. The current flowing in the load may be thought of as the sum of many individual currents flowing in the antenna, induced by the radio wave acting along the length of the antenna. When all the individual induced currents are in phase at the load the maximum amount of power may be extracted from the radio wave. The condition of proper phasing is called *resonance*. Resonance may be established by cutting the length of the antenna to some physical relationship with the size of the intercepted wave. In a simple antenna a resonant condition is usually found at multiples of one-quarter wavelength ($\frac{1}{4}$, $\frac{1}{2}$, $\frac{3}{4}$ wavelength, etc.). When the antenna is operated in an off-resonant condition the sum of the individual induced currents is reduced from the maximum value and the antenna exhibits *reactance* at its load terminals. The ratio of the reactance of the antenna to the radiation resistance is termed the Q of the antenna. When the Q of the antenna is low the reactance is small and varies slowly as the frequency of operation is varied from the resonant frequency of the antenna. An antenna having a high value of Q will tend to be frequency selective and its operating efficiency and energy transfer will tend to be poor when the operating frequency is far removed from the resonant frequency. The antenna Q, therefore, is a measure of response of the antenna in terms of being "sharp" or "broad". In general, the lower Q antenna is more tolerant of adjustment and is easier to place in operation with a minimum of fuss than the higher Q antenna. Either antenna is capable of operation over the relatively narrow segments of the amateur bands.

Operating Bandwidth

The *operating bandwidth* of the antenna is the frequency span over which the antenna performs in an efficient manner. This vague concept must take into account the loss in power gain when the antenna is operated off-frequency, and the increase in SWR (standing wave ratio) on the transmission line under such conditions. The operating bandwidth can therefore mean exactly what the designer wishes it to, having little concept with actual antenna operation.

Many amateur transmitters perform poorly when the SWR on the transmission line approaches a value of 2/1. In some instances, damage to the equipment may happen when attempting operation into antenna systems exhibiting this value of SWR on coaxial transmission lines. For this reason, the operating bandwidth of the antennas described in this Handbook has arbitrarily been established as that frequency excursion that produces a SWR value of 1.75/1 on a coaxial transmission line. The actual operating range of the antenna is usually greater than these arbitrarily defined limits, but the confining factor in most cases is the matching system employed to couple the antenna to the transmission line. It has been the experience of the authors that at SWR values greater than 2/1 the front-to-back ratio of the antenna is usually poor and loading difficulties are often encountered. Operation of the antenna system at high values of SWR are therefore not encouraged.

DIRECTIVITY AND APERTURE

Directivity

In the case of a transmitting antenna, *directivity* is defined as the ability of the antenna to concentrate radiation in a particular direction. All practical antennas exhibit some degree of directivity. A completely nondirectional antenna (one which radiates equally well in all directions) is known as an *isotropic radiator*, and only exists as a mathematical concept. Such a radiator if placed at the center of an imaginary sphere would "illuminate" the inner surface of the sphere uniformly.

Power Gain

Power gain is a term used to express the power increase in the radiated field of one antenna over a standard comparison antenna. The comparison antenna is usually a half-wave dipole having the same polarization as the antenna under consideration. Power gain is measured in the optimum direction of radiation from the antenna.

Effective Aperture

Effective aperture is closely associated with directivity and power gain. In a simplified analogy it may be thought of as the frontal area over which the receiving antenna will extract signal power from the radio wave. Sometimes this concept is referred to as *capture area*. Most high Q arrays, such

The hard work starts at ground level! W8QQ cracks the soil in his backyard in Columbus, Ohio as first step in tower erection for the new Quad antenna array.

as the parasitic beam and the Quad have an effective aperture considerably larger than the physical size of the antenna.

Front-To-Back Ratio

The power gain of an antenna may be thought of as being obtained by taking power radiated from unwanted directions and squirting it out the "front" of the array. Of great interest to the user of a beam antenna is the amount of power that still escapes from the back and sides of the array. This power is wasted, and should be minimized as it contributes nothing to signal gain. The ratio of the power radiated in the forward direction of the antenna as compared to that amount radiated in the opposite direction is termed the *front-to-back ratio* (F/B ratio). Ratios of the order of 5 db to 25 db may be obtained from simple beam antennas, such as the Quad-type. *Front-to-back ratio measurements on amateur antennas will vary widely from these figures as a result of complex wave reflections from the ground and nearby objects.*

The Decibel

Power gain and F/B ratio of beam antennas are usually expressed in terms of *decibels*. The decibel is not a unit of power, but a *ratio* of power levels. In antenna work the decibel may be used as an absolute unit by

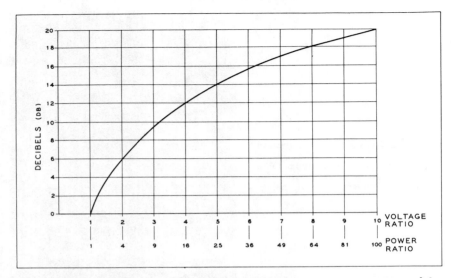

Fig. 1 Power gain and front-to-back ratio of beam antennas are expressed in terms of decibels. Relationship between decibel and voltage or power ratio is shown in this chart. Most receiver S-meters have scale that is calibrated in decibels but which bears no relation to true measurement.

fixing an arbitrary level of reference. If this reference level is taken as the power figure of a dipole antenna, another antenna may be said to have a gain figure expressed in decibels relative to the dipole. This is the reference level used in this Handbook. One decibel unit equals ten times the common logarithm of the power gain over a dipole antenna, as shown in the accompanying illustration (figure 1).

Antenna Gain Measurements

It has been proven on multi-million dollar experimental antenna ranges that data collected at low frequency (below 100 MHz) pertaining to gain, front-to-back ratio, and bandwidth of antennas are decidedly unreliable. Factors such as ground reflection, and the proximity of nearby objects obliterate reference factors and reduce the results to meaningless figures.

Accurate and reproducible figures pertaining to antenna arrays are only determined under exacting conditions. Measurements can be made in the VHF region with model antennas carefully placed so that ground effects and proximity effects of nearby objects are either absent, or are known and computable. The standing wave ratio, bandwidth, and the impedance match of such antennas are carefully controlled, and laboratory equipment is used by trained engineers to obtain results which are then carefully evaluated against the environmental conditions prevailing during the test. Sad to state, the performance figures of antennas tested in this fashion tend to be lower than the figures given by aggressive antenna manufacturers and accepted by hopeful amateurs.

A Reliable Test Procedure

The gain information, dimensional figures, and SWR curves to follow are based upon measurements made in the 144 MHz amateur band. It has been common practice for antenna adjustments and gain figures to be taken with the aid of a field strength meter located at some remote point. Such an arrangement can prove quite helpful when used properly, but any attempts to measure antenna gain or to compare two antennas by measuring field strength along a ground path are subject to extreme inaccuracies. Factors such as ground reflection and the proximity effects of nearby objects obliterate reference points and reduce the results to meaningless figures. Unless the test antenna and field strength meter are located many wavelengths above ground the effect of ground reflection will throw doubt on every measurement. This condition cannot be emphasized too strongly. Many published measurements have been in error due to what is apparently a lack of understanding of the pitfalls produced by ground reflection.

A reliable system of measuring antenna parameters has been developed by the military and industry which utilizes scale models of the antenna under test operating in the very high frequency range. This technique involves plotting the radiation pattern of the antenna and determining the antenna gain from inspection of the pattern. This is done by rotating the antenna under test through 360 degrees and recording the change in relative field strength at a point 10 or more wavelengths away. By running this test in both the E and H planes of the antenna, it is possible to obtain a three dimensional view of the radiation pattern of the antenna which includes the spurious lobes and back radiation. Using this data the actual power gain may be calculated with the aid of formulas (5) and (6), figure 2.

The relationships between power gain, effective aperture, and beam widths may be expressed in a simple nomograph, as shown in figure 3. Antenna gain and aperture figures may be derived from the E and H plane beam widths. The graph assumes that all spurious antenna lobes are reduced 10 decibels or more below the strength of the main lobe. Only the half-power beam widths in the E and H planes are required for this measurement.

It is important that a calibrated field strength meter be used, since an accurate indication of the half-power points of the beam pattern must be

FIGURE 2

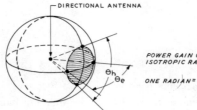

DIRECTIONAL ANTENNA

$$\frac{\text{POWER GAIN OVER}}{\text{ISOTROPIC RADIATOR}} = \frac{\text{SURFACE AREA OF SPHERE}}{\text{AREA OF ELLIPSE AT HALF-POWER ANGLES}} \quad (1)$$

ONE RADIAN = 57.324 DEGREES

THE AREA OF A SPHERE IS EQUAL TO : 4π SQUARE RADIANS (2)

THE AREA OF AN ELLIPSE (OR A CIRCLE) IS EQUAL TO: πAB SQUARE RADIANS (3)

*WHERE A AND B ARE ONE-HALF THE LENGTH
AND WIDTH, RESPECTIVELY, OF THE ELLIPSE
EXPRESSED IN RADIANS.*

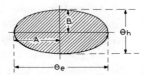

θ_e *AND* θ_h *REPRESENT THE HALF-POWER BEAM WIDTHS IN THE ELECTRIC
AND MAGNETIC PLANES, RESPECTIVELY. THE ELECTRIC PLANE IS GENERATED
IN THE SAME PLANE AS THE RADIATOR ELEMENT, WHILE* θ_h *IS GENERATED
IN THE PERPENDICULAR PLANE.*

$$A \text{ IN RADIANS} = \frac{\theta_e}{114.59} \qquad B \text{ IN RADIANS} = \frac{\theta_h}{114.59} \quad (4)$$

$$\text{THEREFORE: } G = \frac{4\pi}{\pi \frac{\theta_e \; \theta_h}{(114.59)^2}} = \frac{52525}{\theta_e \; \theta_h} = \frac{\text{POWER GAIN OVER}}{\text{ISOTROPIC RADIATOR}} \quad (5)$$

*SINCE A HALF-WAVE DIPOLE HAS A GAIN OF 1.64 OVER AN ISOTROPIC
RADIATOR, THE GAIN OF A DIRECTIONAL ANTENNA OVER A HALF-WAVE
DIPOLE MAY BE EXPRESSED AS:*

$$\text{POWER GAIN (G)} = \frac{52525}{(1.64)\theta_e \; \theta_h} = \frac{32027}{\theta_e \; \theta_h} \quad (6)$$

AN ILLUSTRATION OF HOW POWER GAIN IS THE RATIO BETWEEN THE SURFACE
AREA OF A SPHERE ILLUMINATED BY AN ISOTROPIC RADIATOR AND THE PORTION
OF THE SPHERE WHICH LIES BETWEEN THE HALF-POWER ANGLES, θ_e AND θ_h,
OF THE DIRECTIONAL RADIATOR.

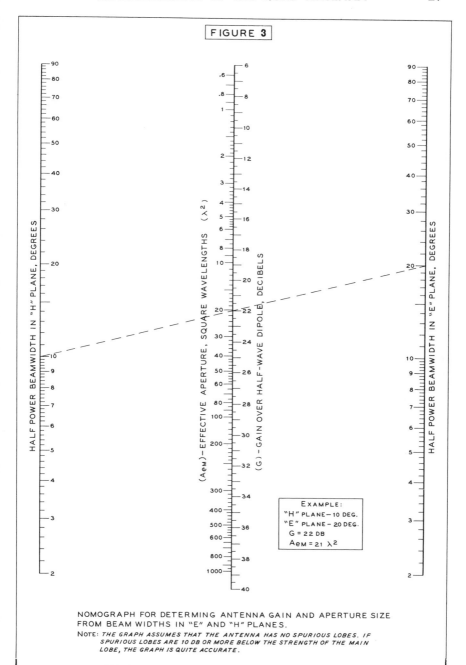

FIGURE 3

EXAMPLE:
"H" PLANE – 10 DEG.
"E" PLANE – 20 DEG.
G = 22 DB
$A_{eM} = 21\ \lambda^2$

NOMOGRAPH FOR DETERMING ANTENNA GAIN AND APERTURE SIZE FROM BEAM WIDTHS IN "E" AND "H" PLANES.

NOTE: *THE GRAPH ASSUMES THAT THE ANTENNA HAS NO SPURIOUS LOBES. IF SPURIOUS LOBES ARE 10 DB OR MORE BELOW THE STRENGTH OF THE MAIN LOBE, THE GRAPH IS QUITE ACCURATE.*

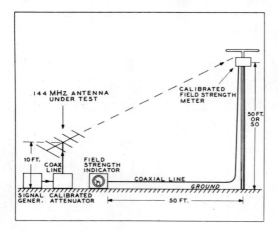

Fig. 4 Antenna measurements can be made with accuracy at 144 MHz. Test antenna is aimed upwards at a distant field strength meter to reduce the effects of ground reflection. Pattern and gain figures are plotted.

made. A good signal generator with an accurate attenuator can be used for calibrating the field strength meter. A typical set-up for measurements of this type is shown in figure 4. The field strength meter is placed atop a fifty foot TV-type "crank up" tower, and the antenna under test is placed a few feet above ground and pointed upwards towards the field strength meter. Ground reflections are thereby reduced to a minimum. The accuracy of readings may be verified by moving the test antenna a few feet and repeating a set of known readings. Any variations in the two test runs indicates that environmental factors are exerting undue influence upon the results and the validity of the tests are open to question. By a trial and error process the optimum placement of the test antenna can be found and a test sequence established that will produce repeatable results regardless of the positioning of the antenna under test. The field strength meter may be passed through the main lobe of the antenna by changing the elevation of the tower.

It was possible for the authors to have some of the tests verified on a commercial antenna range in the Los Angeles area, and the results obtained agreed closely with the information furnished in this Handbook, proving the validity of this method of testing.

Cubical Quad Antenna Parameters

A conventional Quad antenna is the basis of the following data. The configuration of this test antenna is shown in Figure 5. The driven loop is split at the center of the lower segment to provide horizontal polarization.

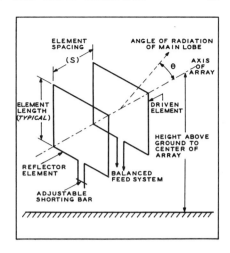

Fig. 5 Typical Quad antenna is composed of two vertical loops spaced a fraction of wavelength. One loop is split at center of the lower segment and is excited with a balanced feed system. Other loop has adjustable shorting bar to permit tuning to the correct parasitic frequency. The sides of loop approximate a free space quarter-wavelength.

The spacing between the two loop elements is adjustable, and the supporting framework is made of wood to eliminate spurious resonance effects in the structure. The parasitic element is broken in the same manner as the driven element permitting adjustment by means of a variable shorting bar. Field strength readings are taken at an extreme distance from the array and the height above ground of the field strength meter is adjustable.

Power Gain and Element Spacing

The overall power gain obtainable with a single parasitic reflector placed behind the driven loop is shown in figure 6. At a spacing of approximately ⅛ wavelength the parasitic element provides an array gain of about 7.3 decibels when adjusted for maximum gain figure. The gain curve is fairly constant for element spacings from 0.1 wavelength to 0.2 wavelength, with the peak of the curve falling near 0.12 wavelength spacing. The reduction of gain at spacings less than 0.1 wavelength is partially due to the loss resistance of the wires in the array. In this particular case the d.c. resistance value was less than 0.1 ohms. At each test point on the curve the parasitic element was tuned for maximum indicated field strength with the result that this curve represents maximum obtainable gain for various element spacings.

Radiation Resistance

The radiation resistance at the center feed-point of the driven element was measured at various spacings and the results plotted in figure 7. A

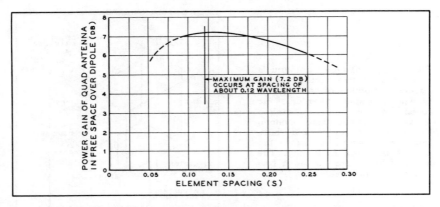

Fig. 6 The gain of a two element Quad array is constant within one decibel for element spacings of 0.08 wavelength to 0.22 wavelength. Maximum gain figure of about 7.4 decibels occurs near the 0.12 wavelength spacing. As the Q of the Quad element is low, the spacing is not a critical factor in the design of the array.

spacing of 0.12 wavelength results in a radiation resistance of about 65 ohms, with the figure rising gradually in value until a radiation resistance of about 140 ohms is reached at 0.25 wavelength. These values can be shifted over a small range by detuning the parasitic element from the point of optimum gain.

As in the case of the simple parasitic beam or other similar antenna, the radiation resistance of the Quad varies with the height above ground. Values plus or minus 15 percent of the curve of figure 7 may be found at different heights above a nominal half-wavelength elevation. Below this elevation figure, the radiation resistance drops to about one half of normal value at a height of one quarter wavelength. Thus, impedance measurements made in a given location for one Quad antenna might not apply to a second antenna situated in a different environment. It would seem, therefore, that impedance measurements might well have to be made on each installation to determine a close value of radiation resistance.

The degree of variation of radiation resistance of the Quad as compared to a typical three element parasitic beam is shown in figure 8. Both the value of radiation resistance and the variation of the radiation resistance are larger for the Quad than for the parasitic beam.

Angle of Radiation

The vertical angle above the horizon of the main lobe of the Quad antenna for various heights above ground is shown in figure 9. It can be seen that (as in the case of the parasitic beam) the angle above the horizon

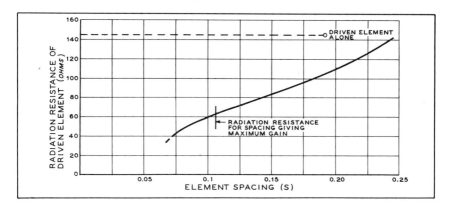

Fig. 7 The radiation resistance of a two element Quad varies between 40 ohms and 140 ohms as element spacing increases from 0.07 wavelength to 0.25 wavelength. Radiation resistance of Quad may be adjusted to 52 ohms by making spacing 0.08 wavelength, or 72 ohms with 0.13 wavelength spacing.

of the lobe of maximum radiation is a function of the height of the array above the ground. *No adjustments made to the antenna (other than changing the height above ground) will influence the angle of radiation of the main lobe.*

ABSENCE OF HIGH ANGLE RADIATION

The term *angle of radiation* of any antenna may be taken to mean the angle above the horizon subtended by the axis of the main lobe of radiation. With practical amateur antennas the radiation lobe is not a knife-edge of energy, nor is it even as sharp as the light beam from an automobile head-

Fig. 8 Radiation resistance of Quad array varies as the height above the ground changes. Measurements are made on Quad having element spacing of 0.15 wavelength. Parasitic array exhibits the same effect but in a much milder form.

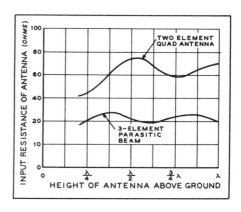

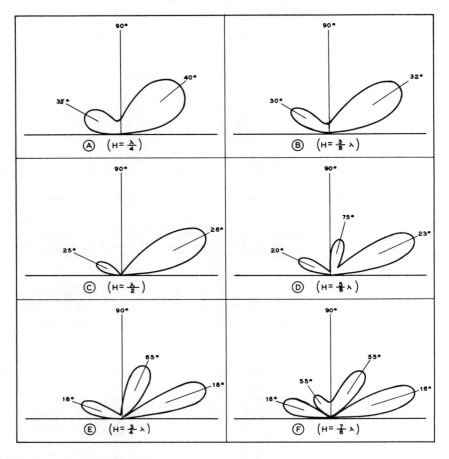

Fig. 9   The vertical radiation pattern for a two element Quad antenna varies in much the same manner as patterns of parasitic beam as the height above ground of the array is changed. Plot A: At a height of one-quarter wavelength angle of maximum radiation of main lobe is 40 degrees compared to 90 degrees for a dipole. As the height of Quad is raised to three-eighths wavelength, the angle of radiation lowers to 32 degrees, as compared to angle of 42 degrees for a dipole. Plot C: Angle of radiation of Quad is 26 degrees at height of about one-half wavelength, four degrees lower than pattern of dipole. Plot D: At height of five-eighths wavelength angle of radiation of Quad is approximately same as that of dipole, but the large vertical lobe of the dipole pattern is suppressed into minor lobe of Quad shown at angle of 75 degrees. Plot E: All lobes of Quad array exhibit lower radiation angle at height of three-quarters wavelength, but the minor lobe grows in size, splitting into two lobes at height of seven-eighths wavelength (Plot F). A parasitic beam has almost the same field patterns as those of Quad array. The actual reflection patterns of practical antennas vary somewhat from these graphs because the earth is not a perfect conductor or reflector, and is never smooth and uniform. Reflection from nearby metallic objects (utility wires, gutter pipes, TV antennas, etc.) tend to distort patterns of any antenna. Even so, these plots emphasize importance of height for desirable low angle of radiation.

Fig. 10 Angle of radiation of main lobe of Quad array is function of the antenna height as measured to lower element. For best DX results Quad should be mounted at least one-half wavelength above surface of ground.

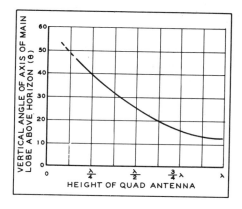

lamp. Rather it is a bulbous lobe, occupying a large area in front of the antenna array.

If the antenna were to be suspended in free space the main lobe of radiation would be directly in line with the aperture of the antenna. When the antenna is located close to the surface of the earth a conflict takes place between the direct wave from the antenna and the wave that is reflected from the surface of the ground. A phase difference occurs between these two waves causing cancellation or reinforcement at various angles above the horizontal. The degree of cancellation or reinforcement is dependent upon the difference in path length between the direct and reflected signals and upon the phase difference caused by the reflection.

Shown in figure 9 are the vertical radiation patterns of a two element Quad antenna for various heights above ground. It can be seen that these patterns bear a resemblance to those of the dipole, except that the high angle radiation present in the case of the dipole is absent in the Quad antenna pattern. This absence of high angle radiation is probably caused by cancellation in the vertical plane by the action of the upper and lower sections of the Quad loop. It is interesting to note that the Quad antenna exhibits approximately the same angle of radiation of the main lobe as the dipole *except* at the lower heights. At a height of 0.5 wavelength, for example, the angle of radiation of the main lobe of the Quad antenna is about four degrees below that of the dipole. At an elevation of ⅜ wavelength the angle of radiation of the Quad is almost ten degrees below that of the dipole. At a height of one-quarter wavelength the dipole is almost useless as a shortwave transmitting antenna since most of the radiation is directed upwards. The Quad antenna, however, maintains its main lobe at an angle of 40 degrees above the horizon *at the same height elevation.* This points up the fact that the Quad will still give satisfactory performance under circumstances where the antenna cannot be elevated well in the air. The same effect is noted with a parasitic beam. The angle of radiation for various heights is shown in figure 10.

Generally speaking, the angle of radiation of the main lobe of the dipole, the parasitic array, and the Quad antenna are identical for all heights above one half wavelength. This belies the claim that the Quad has a lower angle of radiation than other types of antenna arrays.

The Horizontal Pattern of the Quad

The horizontal radiation patterns of the Quad antenna measured at the vertical angles which produce maximum field strength are shown in Figure 11. The shape of the pattern is relatively independent of the height above ground of the array except for variations in the front-to-back ratio which occur because of changes in induced reflector current. The horizontal pattern provided by the two element Quad antenna has a half-power beam width of approximately 62 degrees which is comparable to the beam width of a two element parasitic beam.

It is interesting to note that the Quad has a small amount of vertically polarized energy at right angles to the main lobe of the array. This field is a result of incomplete cancellation of radiation from the two vertical wires of the loop.

Parameter Variations of the Quad

Experience with simple parasitic beams has shown that beam performance quickly deteriorates when the operating frequency of the antenna approaches the resonant frequency of the parasitic element. In the case of the two element beam, the power gain of the array falls off sharply as the operating frequency approaches that of the parasitic element, yet it drops off quite slowly as the operating frequency is shifted away from that of the parasitic. The shape of the gain curve, therefore, will vary depending upon whether the parasitic is a reflector or director (figure 12). This relationship also holds true with the Quad. With an antenna using a reflector element, the power gain of the array drops off quite sharply on the low frequency side of resonance, and drops off comparatively slowly on the high frequency side of the operating frequency. As a result, greater operating bandwidth may be achieved by tuning the Quad reflector for operation at a lower than normal frequency. This process will increase the bandwidth at the expense

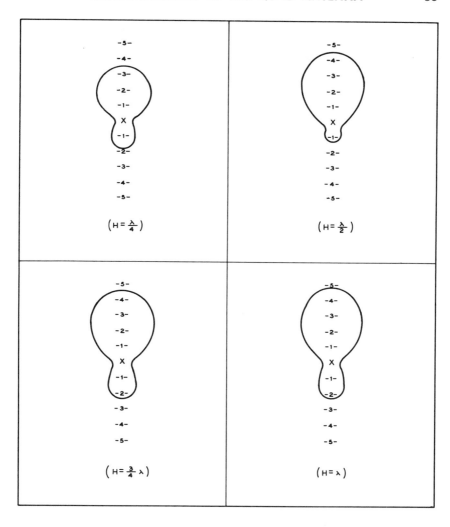

Fig. 11 Shown above is the horizontal field pattern measured at the vertical angles which produce maximum field strength. In this case, the plots were taken at one-quarter, one-half, three-quarters, and one wavelength elevation. Shape of pattern is relatively independent of elevation except for variations in front-to-back ratio which occur because of changes in the induced reflector current. Front-to-back ratio can be optimized at each elevation if the reflector is retuned in each case. The horizontal pattern of a typical 2 element Quad antenna is about 60 degrees wide at half-power points, which is comparable to the pattern of a 3 element parasitic (Yagi) beam. The sharpest pattern is obtained by omitting adjustment stub in the parasitic element and cutting the element to proper size so that no stub is required. Current distribution is thus improved in the parasitic element, and overall operation of the Quad is enhanced, resulting in slightly better gain and front-to-back ratio.

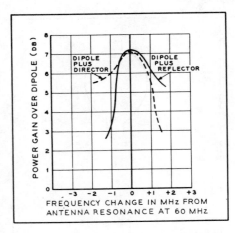

Fig. 12 **As operating frequency of parasitic array is varied the power gain over dipole drops rapidly as the parasitic element becomes resonant. The shape of the gain curve thus depends whether parasitic is a reflector or a director.**

of the maximum power gain. The gain-vs.-frequency characteristic of a typical Quad antenna is shown in figure 13.

F/B Ratio of the Quad

The front-to-back ratio of a properly adjusted Quad reaches a maximum figure at the design frequency and decreases sharply as the operating frequency approaches the self-resonant frequency of the parasitic element. The decrease in F/B ratio is much less severe as the operating frequency is removed from the self-resonant frequency of the parasitic. Maximum F/B ratio at the design frequency is about 25 decibels, dropping to less than 10 decibels when the operating frequency is lowered 3 percent. At 3 per-

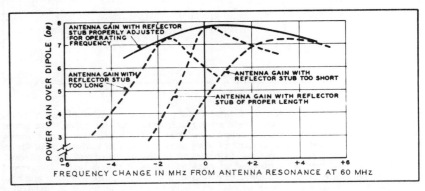

Fig. 13 **When Quad employs parasitic reflector greatest operational bandwidth occurs on high frequency side of resonant point. Quad may operate efficiently plus or minus 3% of design frequency as shown above if the reflector element is retuned for off-frequency operation of the array.**

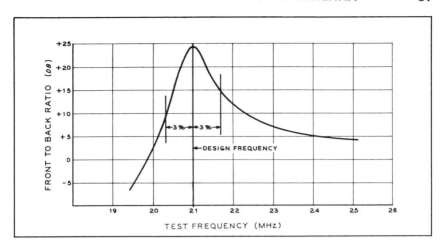

Fig. 14 Front-to-back ratio drops rapidly as the Quad antenna `is operated off-frequency. Even so, Quad qualifies as "broad band" antenna, as it retains good F/B ratio over span of approximately 6% of design frequency.

cent higher than the design frequency the F/B ratio is better, being in the neighborhood of 15 decibels. The F/B ratio of a typical Quad antenna designed for 21 MHz is shown in figure 14.

During this series of tests it was noted that adjustments made to the reflector stub changed the resonant frequency of the driven element of the antenna. In fact, it was possible to tune the array by merely changing the length of the reflector stub. Experiments were conducted to determine the effect upon power gain and F/B ratio of variations in stub tuning. Measurements tended to show that the F/B ratio was critical as to stub tuning and that changing the stub length a small amount would deteriorate the F/B ratio.

The VHF measurements indicated that the power gain and the F/B ratio are dependent upon the length of the reflector stub—at least as much as in the case of the simple parasitic beam. It was also found that unless the reflector stub was of proper length the frequency of maximum F/B ratio was not coincident with the frequency of maximum gain, nor was it the same as the frequency of minimum standing wave ratio on the transmission line. This deviation was most apparent when the reflector stub was too short. In this case (figure 15) the frequency of maximum F/B ratio and minimum SWR are so far apart that the F/B ratio at the frequency of maximum gain is only 10 decibels!

These results emphasize the fact that the maximum F/B ratio of the Quad antenna and the gain of the array can be determined by the length of the

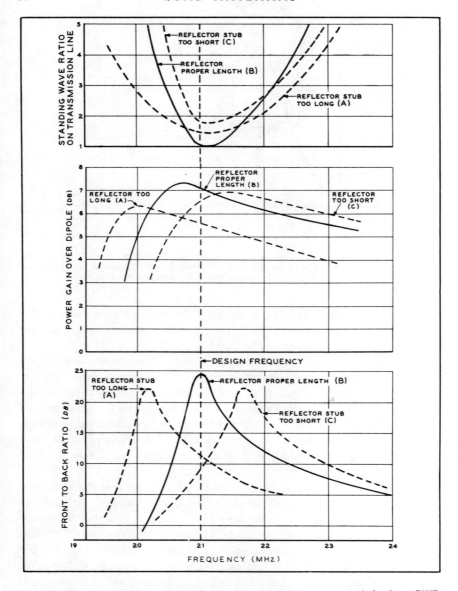

Fig. 15 Frequency of best front-to-back ratio, maximum gain, and the best SWR are not the same unless side length and stub length are adjusted to optimum proportions. Proper side dimensions ensure that optimum F/B ratio, gain, and minimum SWR can be obtained by proper adjustment of the reflector stub. Improper side dimensions produce inferior results, as does the maladjustment of stub. Experimental data proves that proper side dimension of the Quad is very nearly equal to electrical quarter-wavelength in space (Figure 16).

reflector stub *if the dimensions of the sides of the array are properly chosen.* The resonant frequency of the array, however, should be determined by loop dimensions. If the array is the *correct size* to begin with, it is possible to reach the condition of maximum gain by tuning the stub for array resonance at the operating frequency, or for the best F/B ratio.

If the side dimensions of the Quad are incorrect, tuning the stub for maximum F/B ratio or for array resonance will result in a gain figure less than optimum. The important parameter, therefore, is the physical length of the sides of the Quad antenna. Once the correct dimensions have been determined, the antenna will perform in a satisfactory manner with only simple preliminary adjustments.

Quad Antenna Dimensions

The physical length of a half-wave dipole antenna is somewhat less than the length of an electrical half-wave in free space. This length reduction is caused by the so-called "end effects" of the antenna, and also because the antenna is not infinitely thin. A dipole element made of aluminum tubing having a diameter of 1/300 of a half wavelength is reduced in length by a factor of about 0.95. The Quad antenna, on the other hand, may be composed of wire having a very small diameter compared to the wavelength (about 1/2500). In addition, there is no "end" to the wire as the elements of the Quad form a continuous loop. As a result, both of these shortening effects are absent. A lengthening effect is actually present, since the action of bending the wires into a square produces exactly the opposite effect, and the sides of the Quad antenna turn out to be slightly *longer* than a free space quarter wavelength. In actual practice the optimum side length of the Quad is very close to 0.257 electrical wavelength. A table of dimensions for the Quad antenna is given in figure 16.

The Quad Antenna With Parasitic Directors

Quad antennas have been built with one or more director elements, much as in the manner of Yagi antennas. The director elements have sides averaging 5% to 7% smaller than the driven loop, depending upon whether or not tuning stubs are used. Gain figures appreciably higher than Yagi antennas having an equivalent number of elements may be achieved because of the vertical stacking gain of the Quad. Typically, a 3 element Quad will provide a gain of about 9.3 decibels, a 4 element Quad a gain of about 10.2 decibels and a 5 element Quad a gain of about 11.0 decibels (see Figure 6, Chap. IV).

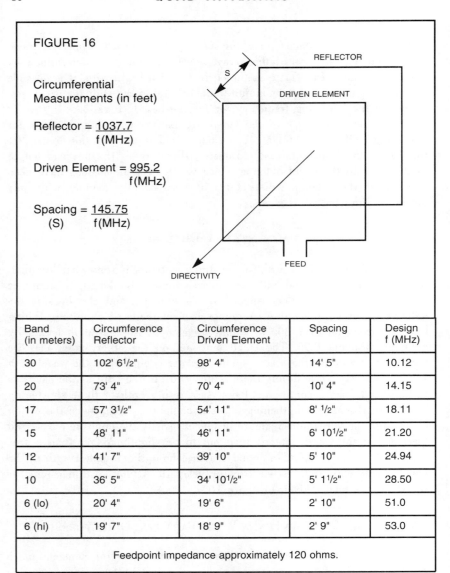

FIGURE 16

Circumferential
Measurements (in feet)

Reflector = $\frac{1037.7}{f(MHz)}$

Driven Element = $\frac{995.2}{f(MHz)}$

Spacing = $\frac{145.75}{f(MHz)}$
(S)

REFLECTOR

DRIVEN ELEMENT

FEED

DIRECTIVITY

Band (in meters)	Circumference Reflector	Circumference Driven Element	Spacing	Design f (MHz)
30	102' 6½"	98' 4"	14' 5"	10.12
20	73' 4"	70' 4"	10' 4"	14.15
17	57' 3½"	54' 11"	8' ½"	18.11
15	48' 11"	46' 11"	6' 10½"	21.20
12	41' 7"	39' 10"	5' 10"	24.94
10	36' 5"	34' 10½"	5' 1½"	28.50
6 (lo)	20' 4"	19' 6"	2' 10"	51.0
6 (hi)	19' 7"	18' 9"	2' 9"	53.0

Feedpoint impedance approximately 120 ohms.

Fig. 16 Dimensions for HF Quads. Number 12 A.W.G. hard-drawn (or pre-stretched) copper wire recommended for 30, 20 and 17 meters. Number 14 A.W.G. for 15, 10 and 6 meter antennas. Cut each loop within plus or minus 1/2-inch for best results. Subtract two inches from length of driven element for inclusion of insulator at feedpoint. Adjust loops on frame to take up slack in wires. Match Quad to 50-ohm line with a quarter-wave section of 75-ohm line. Use of antenna tuner (ATU) at transmitter is suggested.

CHAPTER IV

Multi-element and Concentric
Quad Antennas

It is possible to employ multiple Quad loops to form three element Quad antennas, or to construct an array of concentric Quad antennas capable of operation at several unrelated frequencies. Quad-type arrays may also be formed having "legs" a half-wave in length instead of the usual quarter-wave configuration. The case of the three element Quad antenna will be discussed first.

The Three Element Quad Antenna

The physical configuration of a three element Quad antenna is shown in figure 1. An additional loop is tuned so as to act as a director, and is placed on the opposite side of the driven element from the reflector. Both reflector and director are parasitically excited. Provisions are made for tuning the reflector to a frequency somewhat below the operating frequency by means of a closed stub and for tuning the director to a frequency above the operating frequency by means of an open stub. The open stub is shorter than a quarter-wavelength and exhibits capacitive reactance at points *A-B* on the director loop. Alternatively, the director loop may be tuned with a capacitor at points *A-B* or it might be made smaller in size than the other elements and tuned for optimum gain by means of a shorted stub.

Parameters and Patterns for the Three Element Quad

Test data were derived from a scale model three element Quad antenna operating at a design frequency of 144 MHz. Provisions were made for tuning the parasitic loops and for varying the spacing between the elements of the array. Measurements were conducted as described in Chapter 3.

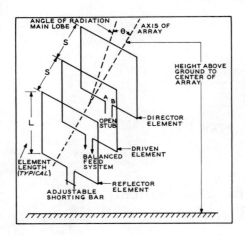

Fig. 1 Three element Quad makes use of open director stub to provide capacitive reactance (see Fig. 9E). However, director may be self-resonant, or undersized with closed stub. See Fig. 6 for typical dimensional data.

Power Gain and Element Spacing

Spacing between the parasitic elements and the driven element of the three element Quad antenna is not critical. The gain-vs.-spacing curve is relatively flat from 0.1—0.25 wavelength, with a slight peak in gain occuring at spacings of about 0.15 wavelength. Employing this spacing, the maximum power gain curve relative to operating frequency is shown in figure 2. At the design frequency a power gain of about 9.3 decibels is obtained. Shown also are the response curves for a two element Quad and a three element Yagi. It can be seen from the curves that the addition of the director element to the two element Quad provides a boost of 2.0 db of power gain at the design frequency. In the case of the simple parasitic "two element beam," the addition of a tuned director to form a "three element beam" produces a boost in power gain of about 2.7 decibels. It is therefore interesting to note that adding an extra parasitic element to the two element Quad provides less additional power gain than adding an extra element to the usual "two element beam." The reason the performance of the extra Quad element is less than optimum is obscure, but it may be due to the fact that the loop-type parasitic element by virtue of its configuration is a low Q design, and it has been demonstrated that high Q parasitic elements are mandatory for maximum signal gain.

Careful antenna gain measurements have shown that the two element Quad closely approaches the power gain of the three element Yagi and the three element Quad surpasses the gain of the three element Yagi by about 1.2 decibels. A check was made of these gain figures on a commercial antenna range and the results agreed closely with experimental data gathered as discussed in chapter III.

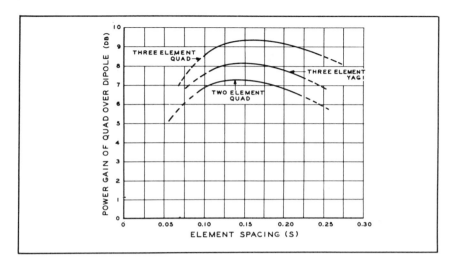

Fig. 2 Three element Quad shows power gain of about 9.3 decibels at element spacing of 0.15 wavelength. Gain is fairly constant for spacings of 0.13 to 0.22 wavelength and averages about 2 decibels better than 2 element Quad.

An additional advantage accruing to the Quad antenna is that it is cheaper to build than an equivalent Yagi, uses no aluminum tubing, has less "wing span" and less wind resistance. It also may be internally stacked to form multi-band arrays. In many countries where aluminum tubing is hard to get or unobtainable, the Quad antenna is the only practical high gain array for most amateurs.

A chart of antenna power gain is shown in figure 3, placing the Yagi and Quad beams in position according to power gain over a dipole antenna. The old rule, "the higher the gain, the louder the signal" still applies to antennas today, and the chart tells the story.

Bandwidth and F/B Ratio

The operating bandwidth of a three element Quad antenna adjusted for maximum power gain is quite limited. The reflector and director are in a near resonant condition and slight frequency excursions will upset the balance of the antenna. The curve of figure 4 shows the bandwidth of this type of array at the 1.75/1 points of SWR is about 300 kilohertz at an operating frequency of 21 MHz. The bandwidth figure expressed in percent of the operating frequency is 1.43%. This means that the SWR will remain less than 1.75/1 over a frequency region less than 1.43% as wide as the

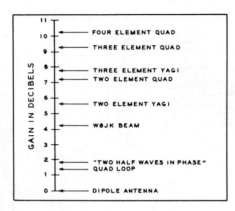

Fig. 3 Gain "ladder" shows the relative power gain in decibels of popular antennas. Two element Quad provides almost as much gain as three element Yagi. The four element Quad is "king of the band". Gain is compared against dipole antenna (0 decibel).

operating frequency. This figure encompasses a 200 kHz span at 14 MHz, a 300 kHz span at 21 MHz and a 400 kHz span at 28 MHz. The SWR curve is not symmetrical, being steeper as the operating frequency approaches the resonant frequency of the parasitic director.

The F/B ratio of the three element Quad antenna is better than 30 decibels at a frequency slightly higher than the design frequency, as shown in figure 5. The F/B ratio drops off sharply as the frequency of operation is removed from the design frequency, but stays better than 20 decibels over the operating bandwidth of the array, as defined by the standing wave ratio on the transmission line.

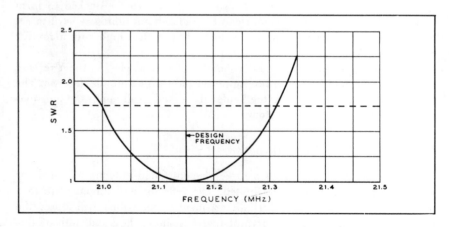

Fig. 4.   Three element Quad array exhibits a narrow bandwidth when elements are tuned for maximum gain and optimum front-to-back ratio. Experimental 21 MHz Quad showed 300 kHz bandwidth over which the SWR on transmission line remained below 1.75/1. SWR increases rapidly on high frequency side of resonant frequency as the parasitic director nears a resonant condition.

"Broad-banding" the Three Element Quad

Greater operational bandwidth may be achieved at a sacrifice in maximum forward gain by employing parasitic elements that are detuned from the lengths of optimum operation. For best broadbanding effect at 10 meters, the reflector stub is lengthened until forward gain drops about one-half decibel and the director stub is lengthened until the gain drops an additional decibel. The power gain under these conditions is approximately 5.5 decibels over a bandwidth of 6%, centered on the design frequency. This bandwidth is sufficient to completely cover the 28 MHz band, or the lower three megahertz of the 50 MHz band. The maximum excursion of SWR on the transmission line at the extremities of bandwidth is slightly under 2.5/1.

The power gain of the broadband Quad compares with that of the two element Quad, but the advantage of the former is that it has a considerably greater operational bandwidth.

Polar Plots of the Three Element Quad Antenna

Measurements of a three element Quad antenna tuned for maximum forward gain were made on a scale model operating at a design frequency of 144 MHz. The polar plots for this antenna are shown in figure 7. An open wire stub was used for the director element, and a closed stub for the reflector. Pattern A indicates that the array is operating at a frequency very close to the resonant frequency of the parasitic reflector, as the F/B ratio is very poor. The SWR on the transmission line is also very high. In pattern B effective reflector action starts to take place and the

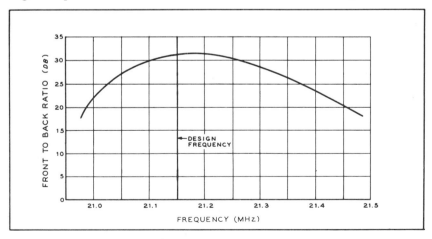

Fig. 5 F/B ratio of three-element 21 MHz Quad is better than 25 db across band.

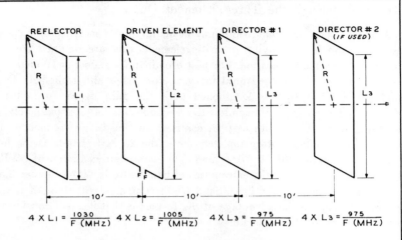

$$4 \times L_1 = \frac{1030}{F \ (MHz)} \qquad 4 \times L_2 = \frac{1005}{F \ (MHz)} \qquad 4 \times L_3 = \frac{975}{F \ (MHz)} \qquad 4 \times L_3 = \frac{975}{F \ (MHz)}$$

FEET X 0.3048 = METERS

DIMENSIONS USING #12 WIRE

Band (in meters)	Circumference Reflector	Circumference Driven Element	Circumference Directors
20	72' 8"	70' 11"	68' 9"
17	56' 10"	55' 6"	53' 10"
15	48' 6"	47' 4"	45' 11"
12	41' 3"	40' 3"	39' 1"
10	36' 0"	35' 1"	34' 0"
6	20' 6"	20' 0"	19' 5"

DISTANCE (R) FROM BOOM CENTER TO LOOP SUPPORT

Band (in meters)	Reflector	Driven Element	Director
20	12' 9 3/4"	12' 5 1/2"	12' 1 1/2"
17	10' 0"	9' 9"	9' 5"
15	8' 6 1/2"	8' 3 1/2"	8' 1"
12	7' 3"	7' 1"	6' 10"
10	6' 4"	6' 2"	6' 0"
6	3' 7"	3' 6"	3' 5"

20M—15M feed with 50-ohm coax;
12M—6M feed with 50-ohm coax and 75-ohm quarter-wave matching section.

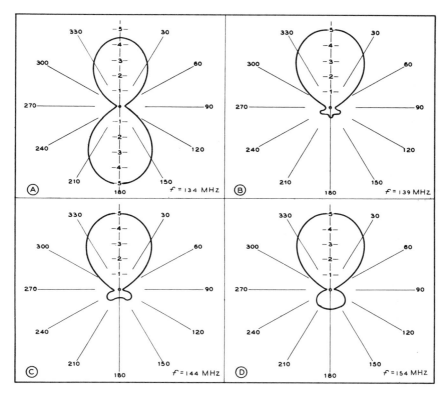

Fig. 7   Polar plots of 3-element Quad show good F/B ratio at design frequency.

rear lobe of the pattern diminishes in size. The SWR at this frequency is approximately 3/1 and the array cannot truly be said to be operating within its operating bandwidth. At the design frequency of 144 MHz (pattern C) the SWR is 1/1 and the beam provides a good pattern, having somewhat better F/B ratio and gain than exhibited by the two element Quad antenna.

As the test frequency is raised, rear lobe splitting takes place and the SWR begins to rise rapidly. The F/B ratio is maximum in figure B (the frequency at which the rear lobe begins to divide) reaching a maximum figure of some 30 decibels. As the frequency of operation is further raised, the SWR climbs rapidly and the rear lobe of the pattern grows in size as the resonant frequency of the director is approached. As shown in pattern D the director begins to act somewhat as a reflector, and a large, distorted radiation lobe appears off the rear of the antenna, accompanied by a large degree of radiation in the plane of the loops.

MULTI-ELEMENT QUAD ANTENNAS

A second director element may be added to the three element Quad to form a four element antenna array. The addition of the second director results in an array gain of about 10.2 decibels as compared to a dipole. This represents a power gain of about 1 decibel or so over the three element Quad and a gain of about 3 decibels over the simple two element Quad. The bandwidth and F/B ratio of the four element array compares favorably with the three element array, the curves being essentially the same.

A five element Quad (three directors, driven element and reflector) was tested at 50 MHz. A power gain of approximately 11 decibels was measured, with an apparent F/B ratio of 30 db. The SWR curve was slightly sharper than that of the four element Quad antenna. A six element test Quad, on the other hand, only provided a power gain 0.5 db better than the five element antenna. The law of diminishing returns seemed to be working at this point.

It can be noted from these figures that the power gain of the Quad array increases slowly as additional director elements are added to the basic two element array. However, because of the relatively low Q of the parasitic loop elements the gain contributed by each loop is less than the value provided by a high Q parasitic element such as found in the normal "parasitic beam." Whereas parasitic beams having twenty or thirty parasitic directors are efficient, high gain antennas, it would seem from these observations that the maximum practical number of parasitic loop elements for the Quad array is limited to four or five. Additional experimental work at a later date might tend to modify this assumption.

Tilt-over tower allows operator to put the finishing touches on 4 element Quad

Concentric Quad Antennas

A popular form of Quad antenna is one wherein various Quad loops for different amateur bands are strung on one framework, as illustrated in figure 8. The most common arrangement is for use on the 20, 15, and 10 meter bands employing three concentric Quads, one for each band. Alternatively, two Quads may be interlaced for two of these bands. In general, results obtained from the concentric Quads compare closely with the operation of widely separated individual arrays, although *interlocking symptoms* may be observed between the concentric antennas. That is to say, adjustments made to one antenna will tend to alter the characteristics of the adjacent antennas. With careful design the results of this unwanted symptom may be reduced to a minimum.

The Two Band Quad

Interaction between the antennas of a two band Quad is quite low. No observable alteration in power gain has been noted when the antennas are interlaced as opposed to isolated operation. The F/B ratio of the *inner* Quad, however, drops a bit as compared to an isolated Quad. F/B ratios of 20 db. or so have been measured on the inner array of a two band Quad, while the outer array held closely to the optimum F/B ratio of 25 db., or better.

A second interlocking effect has been noted concerning the feed system of the antennas. Separate feed systems were used for each Quad under test, consisting of 50 ohm coaxial lines, balancing transformers, and matching networks. The feed systems were adjusted until a 1/1 SWR was obtained

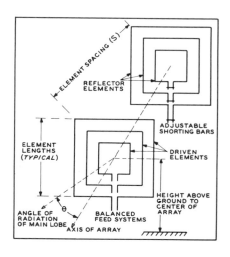

Fig. 8 The concentric or "multi-band" Quad is popular antenna for 20, 15, and 10 meters. When proper dimensions are used, little interaction is noted between the individual Quads. F/B ratio of the middle Quad is lower than that of inner and outer antennas. Balanced feed systems are used with these Quads.

on each Quad antenna at its particular design frequency. Tests were then run for each Quad, plotting the measured value of SWR against the operating frequency of the transmitter. The curves coincided in all respects with those obtained for separate single band antennas. Experiments showed, nevertheless, that the shape and slope of each curve could be varied merely by changing the length of the coaxial line on the *unused* Quad array. Additional tests indicated that the radiation resistance figure of one Quad was also affected by manipulation of the transmission line of the other array.

The effects of interaction were reduced to a minimum by cutting each transmission line to an odd multiple of an electrical quarter-wavelength, the "open" end of the line thereby reflecting a very low terminating impedance across the input terminals of the unused antenna. This, in effect, placed a short circuit across the unused antenna element. Suitable feed systems for Quad antennas which tend to reduce the effects of interlocking to a minimum are discussed in a later chapter.

The Three Band Quad

Interaction between the antennas of a three band Quad may also be observed with adverse effects falling upon the second of the three antennas. In the case of the 20-15-10 meter Quad, this means that the 15 meter section will exhibit inferior F/B ratio when compared to the other two antennas. Effects of interaction upon the transmission lines may be minimized by properly cutting the lines to odd multiples of an electrical quarter-wavelength. The F/B ratio of the second Quad, however, will not be much better than 15 db. in any case and some pickup from the sides of the Quad will also be noted on the 21 MHz band. Gain of the second Quad seems to be comparable to that of the larger and smaller arrays, the deterioration in F/B ratio and growth of side lobes being the only prices that must be paid for the convenience of three band operation with a single structure.

Radiation Resistance of Concentric Quad Antennas

Three Quad antennas for operation on 20-15-10 meters may be interlaced on a single framework having an eight foot boom. Element spacing of the three antennas is such that gain figures for each antenna fall near the peak of the gain curve of figure 6, chapter III. Element spacing is relatively uncritical and may be chosen for a matter of convenience. Eight foot spacing is equivalent to 0.125 wavelength for 20 meters, 0.187 wavelength for 15 meters, and 0.25 wavelength for 10 meters.

Radiation resistance of each separate Quad antenna is a function of the tuning of the reflector stub and can vary over a wide range, depending upon stub adjustment. In general, the value of radiation resistance for a properly tuned Quad is proportional to the element spacing, being highest in the case of the 10 meter section (0.25 wavelength spacing), and lowest in the case of the 20 meter section, with the 15 meter antenna exhibiting a radiation resistance value in between the other two. When tuned for maximum signal gain (coincident with maximum F/B ratio when the proper element dimensions are employed) the radiation resistance of the 20 meter section is about 75 ohms, the radiation resistance of the 15 meter section is about 100 ohms, and that of the 10 meter section is about 120 ohms. Front-to-back ratio drops slowly from optimum value as the element spacing between the front and back sections of the Quad is increased.

Action of the Parasitic Stubs

The parasitic reflector of the Quad array is self-resonant at a frequency lower than the operating frequency of the antenna. Conversely, the parasitic director is self-resonant at a frequency higher than the operating frequency. For optimum results, the parasitic loop should be trimmed to the exact size which determines the correct self-resonant frequency. This may be done with a grid-dip meter. The simpler adjustment technique is to cut the parasitic elements to the same physical size as that of the driven element and then to alter their self-resonant frequency by means of a tuning stub (figure 9).

The reflector element may employ a shorted stub (A) to lower the self-

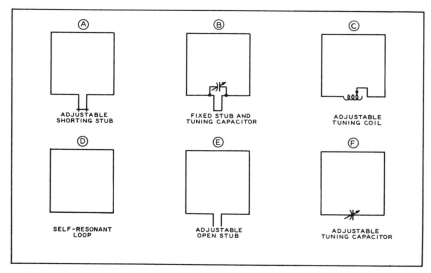

Fig. 9 Parasitic element of Quad may be resonated by shorting stub (A) and (B), or by adjustable coil (C). Optimum adjustment is achieved by making the parasitic element self-resonant (D). Adjustable open stub (E) or series tuning capacitor (F) are sometimes used for stub adjustment.

resonant frequency. Adjustment is made by sliding the shorting bar back and forth along the stub. A fixed stub may be used (B) wherein the electrical length is varied by means of a small capacitor mounted at the top of the stub. The shorted stub may be replaced with a tapped coil (C) which is adjusted turn by turn to reach the correct self-resonant frequency, or a self-resonant loop may be used as shown at D. The self-resonant parasitic element provides slightly higher gain than other configurations, but must be cut to exact size. Typical dimensions for self-resonant reflector loops are given in figure 10, with data for director loops in figure 6.

A director stub is required to electrically shorten the parasitic element, thus raising the self-resonant frequency. An open stub (E) may be employed, or a variable capacitor placed in series with the loop as shown at (F) can be used. In most installations, stubs are to be preferred over coils and capacitors because of the saving in weight and wind resistance.

CONCENTRIC QUAD DIMENSIONS

FIGURE 10 TRIBAND QUAD ANTENNA CIRCUMFERENTIAL MEASUREMENTS				
Band (in meters)	Circumference Reflector	Circumference Driven Element	Spacing	Design f (MHz)
20	72' 8"	70' 8"		14.15
15	49' 0"	47' 0"	8' 6"	21.20
10	37' 4"	34' 8"		28.50
17	56' 9"	55' 2¹/₂"		18.11
12	41' 7"	39' 11"	7' 0"	24.95
10	37' 4"	34' 8"		28.50
10	37' 4"	34' 8"	7' 0"	28.50
6	20' 8¹/₂"	19' 9"		51.0
Approximate feedpoint impedance using #12 wire	20 meters = 70 ohms 15 meters = 100 ohms 10 meters = 130 ohms 6 meters = 150 ohms			

CHAPTER V

The Expanded Quad (X-Q) Antenna

As mentioned earlier in this Handbook, Quad-type arrays may be made up having sides a half-wave in length instead of the usual quarter-wave configuration. A simple example of an antenna of this type is the "Lazy-H" array shown in figure 1A. The gain of this array is about 5.5 db. and is the sum of the gain figures for both horizontal and vertical stacking. The array is fed with a quarter-wave stub coupled to a half-wave phasing section which is transposed for proper phase relationships between the upper and lower bays of the array.

Like the simple Quad, the element tips of the "Lazy-H" array may be bent back upon themselves for feeding purposes and for size reduction (figure 1B). A high degree of field cancellation takes place around the vertical wires and the radiation from these folded portions of the upper and lower sections is thereby diminished. This cancellation effect reduces the gain of the loop from the 5.5 decibel figure of the "Lazy-H" antenna to approximately 5 decibels. This gain is still an impressive figure when compared with the 1.4 decibel gain figure of the quarter wavelength loop employed in the standard Quad antenna.

It is possible to remove the half-wave phasing section from the center of the loop of figure 1B, driving the upper section of the array by connecting the outer tips of the upper and lower sections together as shown in figure 1C. The center of the upper section is left open since the two top wires of the array are out of phase with each other at this point. Addition of a reflector to this expanded half-wave loop produces an *Expanded Quad (X-Q)* array (figure 1D) having an overall gain of about 9.5 decibels as compared against a simple dipole antenna.

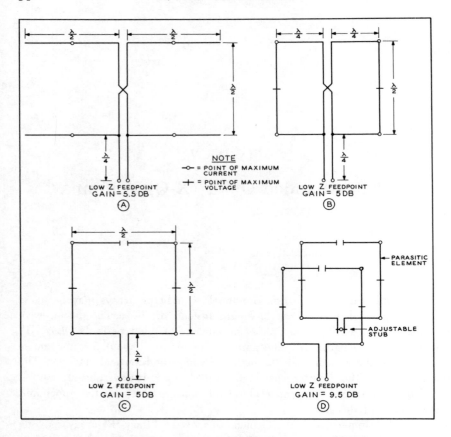

Fig. 1 Expanded Quad (X-Q) antenna is derived from Lazy-H array (A). Ends of "H" are folded back (B) and cross-over feed system is eliminated (C). The X-Q loop provides a power gain of about 5 decibels over a reference dipole. Two element X-Q array provides power gain of 9.5 decibels over a dipole.

DESIGN OF THE X-Q ANTENNA ARRAY

The points of maximum current in the X-Q driven element are shown in figure 1C. For ease of feeding, the antenna wire may be broken at one of these points and driven with a balanced transmission line of the proper impedance. Unfortunately, none of these points fall at the center of the lower section of the array which is a handy place to attach a transmission line. It is possible, however, to connect a *quarter-wave transformer (Q-section)* at the high potential, low current point at the center of the lower portion of the bottom section to provide a convenient low impedance feed point at the base of the transformer. By proper adjustment of this Q-section the X-Q array may be matched to a balanced transmission line having a characteristic impedance in the range of 75-600 ohms.

The X-Q Reflector Element

The reflector loop of the X-Q array is identical to the driven element except that a shorted stub somewhat longer than a quarter-wavelength is used to tune the parasitic element for optimum forward gain (maximum F/B ratio). Because of the use of a tuning stub, it is possible to tune the parasitic element as a director by merely decreasing the length of the stub until it is shorter than a quarter-wavelength. Array gain and F/B ratio are approximately equal for either case.

Maximum array gain occurs with an element spacing of 0.125 wavelength with the gain curve holding to a variation of less than 0.5 decibel for spacings over the range of 0.1-0.25 wavelength. Front-to-back ratios tend to be slightly higher at the closer values of element spacing. A F/B ratio of greater than 22 decibels is obtainable at the element spacing of 0.125 wavelength, which is comparable to the smaller Quad, but the forward radiation lobe is much sharper being approximately 45 degrees wide at the half-power points. As in the case of the simple Quad or the parasitic beam the angle of radiation of the X-Q array above the horizontal is primarily a function of the height of the center of the array above the surface of the ground.

Advantages of the X-Q Array

The X-Q array is easy to construct, requires no expensive aluminum tubing, and provides a power gain figure equal to or slightly greater than a three element parasitic array of normal dimensions. A 10 meter X-Q array is no larger than a simple 20 meter Quad, and the construction of a 15 meter or 20 meter X-Q antenna is not out of the question. Addition of an extra director element to the X-Q array to form a three element expanded Quad has not been tried, but it is not unreasonable to expect a power gain figure of approximately 10 decibels for an array of this type.

Matching the X-Q Array to the Feedline

The impedance at the center of the bottom horizontal section of the X-Q driven element is very high, falling in the range of 2,000-4,500 ohms, the exact value depending upon the size of wire in the array and the physical construction and electrical alignment of the system. A balanced load impedance of this magnitude may be matched to a balanced low impedance transmission line by the use of a *quarter-wave transformer (Q-section)* whose characteristic impedance is the geometric mean between

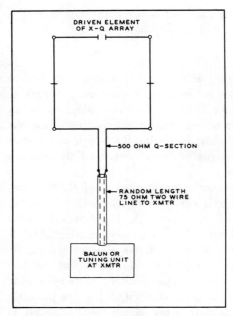

DRIVEN ELEMENT
OF X-Q ARRAY

500 OHM Q-SECTION

RANDOM LENGTH
75 OHM TWO WIRE
LINE TO XMTR

BALUN OR
TUNING UNIT
AT XMTR

Fig. 2 Low impedance points of the X-Q array are located at the corners of the loop making feeding problem more difficult than in the case of the simple Quad. A simple and effective solution is to feed X-Q loop at the center of the lower section (high impedance point) with a quarter-wave matching transformer. This permits a balanced, low impedance transmission line to be coupled to the antenna. Balun or tuning unit is used at transmitter (unbalanced) for pi-network output circuits.

the two impedances that are to be matched. If the Q-section has an impedance of Z_Q ohms and is terminated by a load of Z_L ohms, the impedance reflected to the opposite end of the Q-section is Z_I ohms and is defined by this equation:

$$Z_I = \frac{Z_Q^2}{Z_L}$$

As a practical example, a 75 ohm balanced transmission line (Z_I) may be matched to an antenna (Z_L) whose impedance is 3,300 ohms by using a Q-section having a characteristic impedance of:

Z_I = 75 OHMS	$Z_Q = \sqrt{Z_I \times Z_L}$	$Z_Q = \sqrt{247,500}$
Z_L = 3,300 OHMS		
Z_Q = ?	$Z_Q = \sqrt{3,300 \times 75}$	Z_Q = 498 OHMS

At the resonant frequency of this antenna a 500 ohm Q-section will provide almost a perfect 1/1 standing wave ratio on a balanced two wire 75 ohm transmission line (figure 2). Adjustments to the impedance of the Q-section will permit balanced lines of any reasonable impedance to be used with the X-Q array. Data for designing Q-sections capable of use with the X-Q can be found in the *"Radio Handbook"* distributed by *Editors And Engineers* division, Howard W. Sams & Co., Inc., Indianapolis, IN 46268, and available at large radio distribution houses, radio mail order houses and libraries.

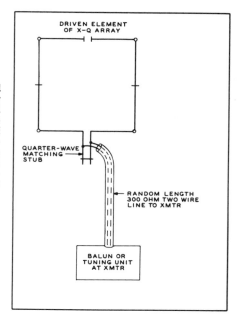

Fig. 3 X-Q array may be matched to low impedance, balanced line by means of a quarter-wave stub tuned to the operating frequency of the antenna. Stub is resonated by use of dip-oscillator. Random length transmission line is tapped onto the stub at 300 ohm feed point, found with aid of dip-oscillator or SWR meter. Shorting bar may require minor adjustment after line is attached to stub.

Matching Stub System

A second matching system making use of a *quarter-wave matching stub* may be used with the X-Q array (figure 3). The array and stub are resonated to the operating frequency by sliding the shorting bar up and down the stub. A dip-oscillator is coupled to the stub to provide a convenient indication of the resonant frequency of the array. Once the correct adjustment has been found the shorting bar is soldered in place. The next step is to determine which point on the stub will match the 300 ohm impedance of the transmission line. Low impedance points will be found close to the shorting bar and higher impedance points are found a corresponding distance up the tuned stub. A dip-oscillator and Antennascope may be employed to find the desired impedance point on the stub. Construction and operating information for the Antennascope may be found in the previously mentioned *"Radio Handbook."*

Coaxial Feed Systems for the X-Q Array

Coaxial feed systems may be employed with the X-Q array as shown in figure 4. A half-wave balun transformer can be used to provide a balanced termination point of 208 ohms (A-B, figure 4). A Q-section having a characteristic impedance of approximately 800 ohms will provide a good match between the balun and the X-Q array. On the other hand, the balun

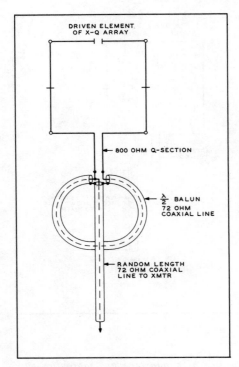

Fig. 4 Q-section, half-wave balun and coaxial line provide unbalanced feed system for pi-network transmitters. Balun provides 208 ohm termination point for 72 ohm line, and adjustable Q-section steps impedance up to several thousand ohms, suitable for high impedance feed point of X-Q.

may be attached to the 208 ohm point on a tuned stub in the same manner as the balanced two wire transmission line.

The X-Q Array Adjustment Procedure

A drawing of the X-Q array giving all important dimensions is shown in figure 5. This antenna is adjusted in much the same manner as the smaller Quad. The relationship between F/B ratio and power gain are set by employing the correct side dimensions and element spacing during construction. The remaining corrections necessary after the X-Q array is erected are resonating adjustments to be made to the parasitic reflector and driven element. The first step is to attach a 75 ohm balanced line and 500 ohm Q-section to the driven loop of the X-Q array and run the line to your receiver. The X-Q array is placed in operating position and the reflector (back of the array) is aimed at a nearby transmitter that has a horizontally polarized antenna. The reflector stub of the X-Q array is now adjusted for minimum signal pickup as read on the S-meter of the receiver. This adjustment should be repeated with several local signals. Once the correct point has been found for the shorting bar on the reflector stub it should be soldered in position.

The next step is to adjust the Q-section of the driven element. The 75 ohm feedline is removed and replaced with a shorting bar. The dip-oscillator is coupled to the bar which is moved up and down the stub an inch or so at a time until the resonant frequency of the driven element falls at the chosen design frequency of the array. During this operation it may be possible to observe a secondary indication of resonance occuring somewhat lower in frequency than that of the driven element. This is the resonant frequency of the parasitic element and should be approximately 3% to 5% lower in frequency than the resonant frequency of the driven element. When the driven element has been set to the proper operating frequency the shorting bar may be: 1) soldered in position to form a matching stub, or: 2) removed and replaced with a low impedance transmission line, thus changing the matching stub into a Q-section.

FIGURE 5			DIMENSIONS FOR X—Q ANTENNA					
BAND	L		SPACING		DIRECTOR STUB		REFLECTOR STUB	
	FEET	METERS	FEET	METERS	FEET	METERS	FEET	METERS
40	66' 6"	20.27	17' 0"	5.18	32' 0"	9.75	37' 8"	11.43
30	48' 7"	14.82	11' 8"	3.55	22' 9"	6.93	26' 7"	8.12
20	33' 5"	10.20	8' 4"	2.54	15' 11"	4.86	18' 9"	5.72
17	27' 2"	8.28	6' 6"	1.99	12' 5"	3.79	14' 7"	4.46
15	22' 3"	6.78	5' 8"	1.74	10' 7"	3.23	12' 8"	3.81
12	19' 9"	6.01	4' 9"	1.44	9' 0"	2.75	10'7"	3.24
10	16' 6"	5.03	4' 3"	1.30	7' 10"	2.39	9' 3"	2.82
6	9' 4	2.84	2' 5"	0.75	4' 5"	1.36	5' 3"	1.60

ANTENNA GAIN = 9.5 DB
F/B RATIO = 22 DB

INSULATORS L

INSULATORS

L

STUB SPACING IS 3 INCHES

FEEDPOINT
(SEE TEXT)

CHAPTER VI

Feed Systems for Quad Antennas

A *transmission line* is required to transmit or guide electrical energy from the transmitter to any antenna, or from the antenna to the receiver. In most cases, some sort of *matching system* must be placed between the transmission line and the antenna to provide an efficient transfer of energy. The reader is referred to the *"Beam Antenna Handbook,"* published by *Radio Publications Inc.,* for a full discussion of transmission lines and antenna matching techniques. The specific case of matching balanced and unbalanced transmission lines to various forms of Quad antennas will be discussed in this chapter.

The Balanced Quad Antenna

The simple Quad driven element is a quarter-wave loop, open at the center of the bottom section for feeding purposes. The loop is symmetrical and the current distribution on the wire is also symmetrical, as shown in figure 1A. The current is a minimum value at the centers of the vertical sides and reaches a maximum figure at the centers of the horizontal sections. If the Quad loop is broken at point X and straightened out into two horizontal wires the current distribution in these wires would appear as shown in figure 1B. The current distribution in the lower wire resembles that of the simple dipole antenna, being a maximum at the center and minimum at the ends of the wire. This wire may be fed at the center with a *balanced* two wire transmission line connected to point A-B. The amplitude of the r-f current in one leg of the line will be equal to the amplitude of the current in the other leg, and 180° out of phase with it. Equal, out-of-phase currents in the wires of the balanced transmission line are of prime importance because the Quad element is a closed loop (unlike the simple dipole) and the current flowing at point A has to equal the current

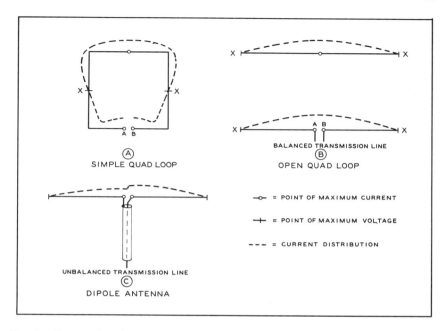

Fig. 1 Current distribution is symmetrical in Quad antenna as it forms a closed system (A). Unequal and improperly phased currents may flow in a dipole antenna (C) as no electrical connection exists between dipole halves.

flowing at point B as these two points are electrically connected together by the wire of the loop. On the other hand, unequal and improperly phased currents may flow in the simple dipole (figure 1C) as there is no electrical connection between the halves of the antenna. A situation such as this arises when the balanced dipole is fed with an unbalanced (coaxial) transmission line.

Transmission Line Radiation

When an unbalanced transmission line is employed to transfer power to a balanced antenna a certain proportion of the line current flows on the outer surface of the coaxial shield. Under proper operating conditions, no electric or magnetic fields extend outside of the outer conductor of the line. All fields exist in the space between the center conductor and the shield. Thus the coaxial cable is a perfectly shielded line. When current flows on the outer surface of the shield the shielding function of the line is lost, as this current is not balanced with respect to the current flowing on the inner conductor of the line. As a result considerable power may be radiated directly from the line. This power does not reach the antenna and is lost for all practical purposes. The field of radiation of the line bears no relationship to the antenna field and usually results in a deterioration of the front-to-back ratio and power gain of the antenna. Complaints that the Quad antenna exhibits no F/B ratio can usually be traced to an unbalanced feed system in which the transmission line is coupled to the antenna in some manner so as to alter the radiation pattern of the antenna. The use of some form of coupling transformer *(balun)* between the feed line and the antenna or the use of a balanced line are two solutions to this.

In most amateur antenna installations the transmission line drops downward from the antenna and under conditions of line radiation may be compared to a long vertical antenna having a high angle of radiation. Since low angle radiation is required for effective antenna performance the field about the transmission line serves no useful purpose at all and only wastes power that otherwise might make the signal stronger and more readable at some distant point of reception.

The first rule, therefore, for the design of an efficient feed system for a balanced antenna such as the Quad is:

1—*The transmission line system must deliver balanced, out-of-phase power to the balanced feed points of the driven element of the array.*

Balanced Feed Systems

A balanced 75 ohm two-wire line may be used to feed the Quad antenna. As the radiation resistance of the Quad and the impedance of the line are not too far apart in absolute value, the standing wave ratio on the line will be reasonably low. If the line is removed from the immediate vicinity of metal objects and the ground, if it does not run parallel to the antenna elements, and when a proper coupling circuit is used at the station, the line currents will be balanced and radiation from the line will be at a minimum. The balanced line may be connected to the transmitter by a simple antenna coupler, such as shown in Figure 2. Heavy duty transmitting type twin-lead is recommended for power levels up to the maximum legal limit.

When the balanced line is cut to multiples of an electrical half-wavelength the line will resemble a series of transformer sections each having a 1-to-1 transformation ratio, reflecting the antenna terminating impedance to the input end of the line. An antenna coupler can then transform this value to a nominal value of 50 ohms, suitable for an unbalanced coaxial output system such as used in the majority of modern transmitters. Antenna coupler taps and tuning are adjusted for lowest value of SWR on the 50 ohm line to the transmitter. Altering the length of the twin line to the antenna a foot or so may help if difficulty is encountered in loading the transmitter properly while maintaining a low value of SWR on the coaxial line.

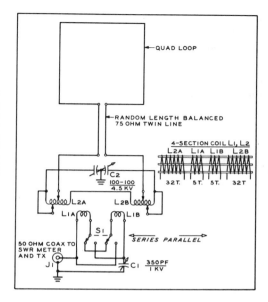

Fig. 2 Antenna coupler for use with balanced line. The coil is 2½" diameter, 8 turns per inch of #14 wire. End sections are shorted at 4 turns for 80 meters, 16 turns for 40 meters, 28 turns for 20 meters, 29 turns for 15 meters and 30 turns for 10 meters. See "Radio Handbook" for more data.

The Concentric Tri-Band Quad

A balanced feed system may be used with a tri-band Quad element as shown in figure 3. The loops are constructed to the dimensions given in Chapter IV, figure 10. The feed points of the loops are then connected in parallel. Using standard dimensions and reflector spacing, the impedance range presented at the feedpoint of the loops lies between 75 ohms and 140 ohms. The loops not resonant at the operating frequency present a rather high impedance across the loop

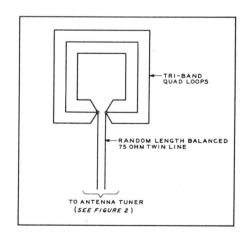

Fig. 3 Tri-band Quad loops may be connected in parallel and fed with random length 75 ohm ribbon line to antenna tuner. If a relatively high value of SWR is accepted, and line radiation not a problem, parallel loops may be fed directly with 50 ohm coaxial transmission line.

in use and each loop exerts a measureable detuning effect upon its companions. As a result, the SWR on the twin-line system at the resonant frequency of the antenna on each band is somewhat higher than in the case of the single band array shown in figure 2. Typically, the SWR on the twin-line feeder of the tri-band Quad will run less than 2-to-1 at resonance. The use of the antenna tuner, however, will drop the SWR on the coaxial line to the station equipment to unity. This simple system is very effective and when used with an antenna tuner will provide good results with a tri-band Quad antenna.

Unbalanced (Coaxial) Feed Systems for the Quad Antenna

In many instances it is convenient or necessary to feed the balanced Quad antenna with a coaxial line having a single conductor which is unbalanced to ground. The line should not be directly connected to the driven element of the Quad, or a severe discontinuity will occur in the electrical character- istic of the transmission system. This will create a high value of SWR on

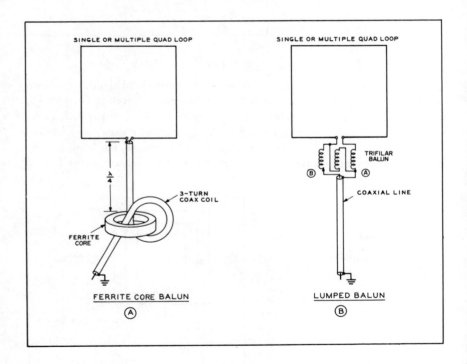

Fig. 4 Quad loop is fed from coaxial line and balun. Ferrite core (A) is Indiana General CF-123 (Q-1 material), 2.4" outside diameter. Information may be obtained from Indiana General Corp., Crow Mills Rd., Keasby, NJ 08832. Trifilar air core, coil balun (B) is shown in Figure 5.

the line and a loss of considerable energy by radiation will take place as a result of unbalanced line currents. No amount of adjustment to the antenna can completely remove the SWR on the transmission line created by this type of discontinuity. In addition, the current flowing on the outer surface of the coaxial shield will lead to erroneous SWR measurements when a simple SWR directional coupler is employed to examine the condition of the transmission line. In order to effect an efficient junction between the unbalanced transmission line and the balanced antenna system a line-balance converter *(balun)* must be used (figure 6). The outer surface of the shield of a coaxial line is normally at ground potential, whereas the inner conductor is well above ground potential. Both conductors of the Quad driven element display the same potential to ground under ideal conditions. The object of the line-balance converter is to produce a high impedance to ground between the outer surface of the outer conductor of the coaxial line at the point where it connects to one terminal of the balanced antenna, thereby converting the end of the coaxial line to a balanced condition.

Two Practical Baluns for Your Quad

Two practical baluns that will do the job are shown in figures 4 and 5. Drawing A shows a r-f choke balun made of three turns of your coaxial transmission line wound about a small ferrite core. The choke is located about a quarter-wavelength down the line from the antenna and the portion of the line between the choke and the antenna forms a simple balun. The outer shield of the line from the choke to the antenna has r-f energy on it and this portion of the line should be brought away at right angles to the antenna wires. Placement of the choke coil along the transmission line is not critical.

The line is wound through the ferrite core which is taped into position. The coil must be fairly large as coaxial line should not be bent around too sharp a radius. A coil diameter of not less than 7 inches is suggested for RG-8A/U coaxial line, and not less than 4 inches for RG-58/U line.

A lumped constant balun may also be used (figure 4B). Two suitable designs are shown in figure 5. At the left is an air core balun having an average power capability of better than 1000 watts over the range of 7 to 30 MHz. The balun has a 1-to-1 ratio and provides good balance to either a 50 or 75 ohm transmission line. The unit consists of three coils of #14 *Formvar* insulated wire, ten turns to each coil. *Formvar* (polyvinyl formal-phenolic resin) is superior to enamel insulation because of its greater dielectric breakdown strength. The windings are placed on a 4-inch long piece of 1-1/16 inch outside diameter gray polyvinyl-chloride (PVC) plastic tubing, commonly used in many areas for water pipe.

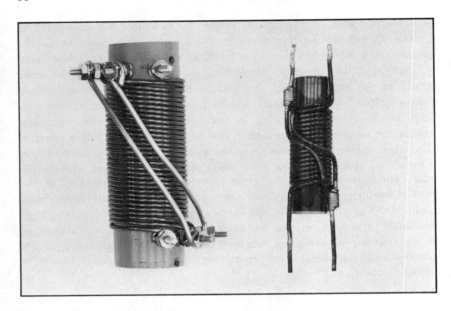

Fig. 5 Air core balun at left is good for 1000 watts PEP power level. Jumpers connect windings in proper sequence. Coil termination (A) is at lower right with termination (B) at lower left. Ferrite core balun is at right.

Three pieces of wire about 4 feet long are needed. The wires are placed parallel to one another and the far ends held in a vise. The near ends are scraped clean of insulation and wrapped around three 4-40 bolts placed in the PVC form as anchor points. The three wires are then wound side by side on the form as one, until ten trifilar turns are on the form. Wind under tension so that the coils adhere tightly to the form. The other ends of the windings are now scraped clean and attached to the respective anchor bolts, as shown in the photograph.

The last step is to interconnect the center, or balancing, winding. The coil is cross-connected across the outer coils at the ends by means of two short straps, the terminals reversed in physical position from one end of the coil to the other.

The input terminals of the balun are non-symmetrical. *Point A* must be taken as ground and is connected to the shield of the coaxial line. *Point B* is connected to the inner conductor of the line. At the output end of the balun, the terminals are symmetrical and balanced to ground.

A compact ferrite core balun is shown at the right of figure 5. It is useable over the range of 3.5 to 30 MHz. The average power capacity is 700 watts up to 14 MHz and 400 watts at 30 MHz. With intermittent voice SSB operation, the power capacity probably can be doubled with safety. The balun is wound on a ½-inch diameter, Q-1 material ferrite rod having a permeability of 125 at 1 mc. A suitable rod is the *Indiana General CF-503*, which is 7½ inches long. Information about this material can be obtained from Indiana General Corp., Crow Mills Rd., Keasby, NJ 08832. The inexpensive ferrite rod can be easily nicked with a file around the circumference at the desired length and broken with a sharp blow.

The balun winding consists of six turns of #14 *Formvar* wire closewound on the rod as described for the air core balun. When wound, the leads to the coils may be wrapped with string and the ends given a coat of epoxy resin. Keep the coil itself free of resin or other material, as the distributed capacitance of the winding must be held to a minimum for proper operation.

Weatherproofing the Balun

Either type of balun must be protected from the weather without upsetting the electrical characteristics of the device. The balun may be placed in a cylindrical case made from a section of a polyethylene "squeeze bottle" such as holds hair shampoo. The ends of the bottle are cut off and plywood discs are substituted, held in place with very small wood screws through the bottle wall. The balun is suspended inside the bottle section by its leads which are connected to brass bolts passed through the plywood discs. When completed, the end discs are given a coat of epoxy resin to waterproof the joints.

An Inexpensive Linear Balun

The purpose of a balun is to decouple the outside shield of the transmission line from the effects of the antenna. An inexpensive linear balun may be used.

The balun is made of flexible, metallic braided sleeving which is cut to length and slipped over the jacket of the coaxial line. The "top" (or antenna) end of the braid is terminated about an inch below the end of the coaxial line and is firmly taped in place. No connection is made between the balun and the coaxial line at this point. The braid is now smoothed down along the line and trimmed to the correct length. To hold it in position it is necessary to wrap it with a few turns of vinyl tape every six inches or so. The "bottom" end of the balun is tinned with a soldering iron and a short length of wire is soldered to the bottom of the braid before the end is taped.

The last step is to remove the vinyl jacket from the coaxial line about ½-inch below the balun, exposing about ½-inch of the flexible outer shield of the line. The wire from the balun is trimmed short and soldered to the shield of the line. The connection is wrapped with vinyl tape to prevent moisture from entering the line. Construction and installation of this simple balun is straightforward. When completed, the braid should be wrapped with waterproof vinyl tape to reduce the effects of moisture in the atmosphere.

Pi-Network Operation

Generally speaking, almost all pi-network circuits employed in modern transmitters will operate into nonreactive loads within the range of 50 to 150 ohms. The nature of the pi-network is such that as the external load impedance is lowered additional output capacity must be added to the network output section. A practical limit is reached in the neighborhood of 20 ohms or so, below which the value of output capacitance required to establish an impedance match between the amplifier tubes of the transmitter and the external load becomes inordinately large. The reverse is true, however, when the pi-network is called upon to match the transmitter to transmission line impedances greater than 50 ohms. In this case a smaller than normal value of output capacitance in the network is required.

In any case, Quad antennas fed with balanced transmission lines may be coupled to pi-network circuits by the use of an auxiliary tuning unit and SWR meter.

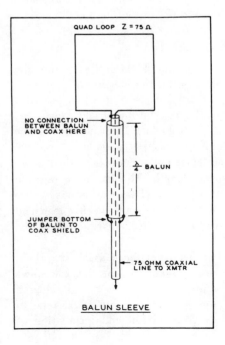

QUAD LOOP   Z = 75 Ω

NO CONNECTION
BETWEEN BALUN
AND COAX HERE

$\frac{\lambda}{4}$ BALUN

JUMPER BOTTOM
OF BALUN TO
COAX SHIELD

75 OHM COAXIAL
LINE TO XMTR

BALUN SLEEVE

Fig. 6 Unbalanced coaxial line may be attached directly to Quad loop with aid of balun sleeve placed at top end of line. Sleeve and outer conductor of line form a quarter-wavelength transformer having a high impedance across the open end (top). Both terminals of the coaxial line are isolated from ground.

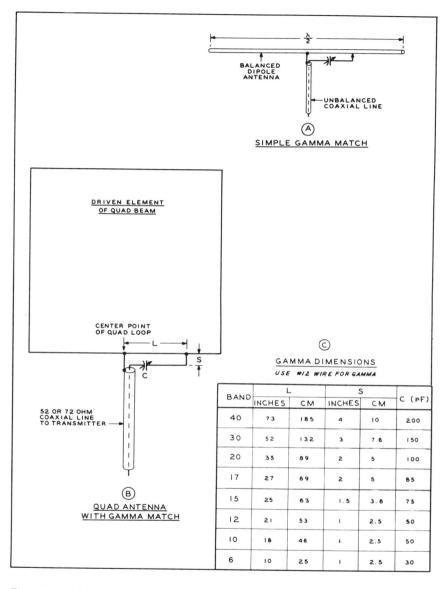

GAMMA DIMENSIONS

USE #12 WIRE FOR GAMMA

BAND	L		S		C (pF)
	INCHES	CM	INCHES	CM	
40	73	185	4	10	200
30	52	132	3	7.6	150
20	35	89	2	5	100
17	27	69	2	5	85
15	25	63	1.5	3.8	75
12	21	53	1	2.5	50
10	18	46	1	2.5	50
6	10	25	1	2.5	30

Fig. 7 Popular gamma match may easily be employed with Quad antenna. Gamma is used to match unbalanced coaxial line to balanced antenna system (A). Same configuration is used with Quad, except gamma is made of wire, and spacing between gamma wire and Quad loop is quite small. Gamma dimensions for all bands are given in chart C. Gamma capacitor setting will be approximately 90% of value shown. For power up to one kilowatt, small wide-spaced receiving type capacitors may be employed.

THE GAMMA MATCH

The *Gamma Match* is a linear transformer capable of matching a low impedance unbalanced transmission line to a high impedance point along a dipole or other driven element. The reactance of the transformer is tuned out by means of a series capacitor. The gamma matching system is a high-Q network and should be adjusted at the resonant frequency of the antenna for best operation. A similar transformer for a balanced feed system is termed a *T-match*.

The *Gamma Match* system has proven to be very effective when employed with parasitic arrays constructed of aluminum tubing (figure 7A). It is used to match unbalanced coaxial lines to either a balanced or unbalanced driven element. The gamma match consists of a single gamma rod running parallel to the driven element and connected to it a short distance from a current loop on the element. A variable capacitor is used to resonate the gamma rod to the operating frequency of the antenna. The matching system may be compared to an auto-transformer having a series tuned input circuit.

The gamma match is constructed of aluminum tubing or heavy wire when used with a parasitic array having tubing elements. The length of the gamma rod, spacing, and size of the series capacitor are a function of the impedance transformation ratio and of the physical diameters of the driven element and the gamma rod.

The gamma match may be applied to the Quad antenna if the dimensions are adjusted to compensate for the thin wire elements of the Quad and the particular value of radiation resistance and operating Q of the array (figure 7B). In this case, the gamma section is made of relatively thin wire instead of tubing. As the wire diameter is small, the spacing of the gamma must be reduced to a few inches in order to provide a proper impedance transformation. The matching system may be used to match any Quad antenna to a coaxial line having an impedance value between 50 and 95 ohms, regardless of the actual radiation resistance of the antenna Since the impedance transformation is continuously variable within these limits the complicated matching stubs, Q-sections, and high impedance transmission lines are no longer required to do the job. By merely adjusting the length of the gamma wire and the setting of the gamma capacitor a close match may be accomplished between the low impedance coaxial transmission line and the driven loop of the Quad antenna. Proper dimensions for gamma matching systems for various Quad antennas are given in figure 7C. These dimensions apply to all forms of Quads, regardless of the number of parasitic elements in the antenna. Generally speaking, higher impedance transmission lines require longer gamma wires than do lower impedance lines. In any case, it is important to keep the gamma wire-antenna spacing to two inches or less for optimum results.

Adjusting the Gamma Match

There are several methods of adjusting the gamma match for proper operation. The purpose of these adjustments is always the same—to resonate the gamma system to the frequency of the antenna and to provide the proper impedance transfer to achieve a 1.0/1 standing wave ratio on the transmission line at the frequency of resonance. Gamma resonance is determined by the length and spacing of the gamma wire and by the setting of the variable capacitor. Proper impedance transformation is determined only by the length and spacing of the gamma wire. Since these two adjustments involve the gamma wire, they tend to be interlocking. Unless the experimenter starts with the system in a near-adjusted state he is apt to go around in circles, compensating for one state of misadjustment by varying the parameters that control the other variable. Before any adjustments are made to the matching system, all dimensions should be set to those given in figure 7C.

Shown in figure 8A is a simple adjustment setup that requires a minimum of equipment. Your transmitter serves as a signal generator and a SWR meter (sometimes called a "reflected power meter", SWR "bridge", or "monimatch") is placed in the coaxial line leading to the antenna under test. As it is necessary to make the tuning adjustments at the antenna, the SWR meter should be placed at the antenna end of the transmission line so that the reading of the instrument can be easily observed by the operator making the tuning adjustments.

The transmitter is tuned to the resonant frequency of the antenna (usually near the center of the amateur band) and is run at reduced power. A SWR reading for this frequency is made and noted on a piece of paper. The

UP IT GOES! The tower is installed and the Quad is complete! W8QQ and his crew are ready to raise the 14 MHz antenna to the top of the tower. In a few minutes they will try the first "CQ" to see if the new beam works.

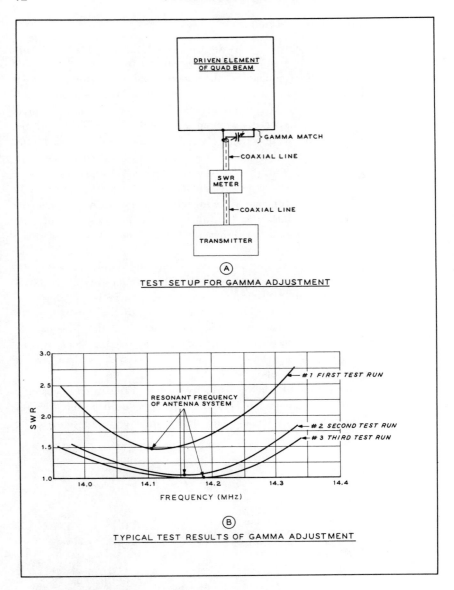

Fig. 8   Your transmitter and SWR meter are only tools required for adjustment of gamma match. Length of gamma and capacitor setting are varied to produce lowest value of SWR at resonant frequency of antenna. First test run (curve #1 of sketch B) shows system is out of adjustment as minimum value of SWR is high and bandwidth is narrow. Subsequent adjustments result in curve #2 which is near optimum. Additional "touch-up" of gamma system produces curve #3 as final result. SWR is now at minimum value and bandwidth is excellent.

gamma capacitor is now varied a few degrees at a time and the new value of SWR read on the meter is noted for each setting of the gamma capacitor. The capacitor should finally be reset to that particular value giving the lowest SWR reading. The gamma wire should now be varied an inch or two in length while notations are made of the change in SWR reading. The length of the wire should be noted for each measurement. When these two series of measurements are completed, the capacitor and wire length are set to provide the minimum SWR readings noted during the tests.

The next step is to leave these adjustments alone for a moment and log SWR readings as the transmitter is tuned back and forth across the amateur band. Readings should be taken every 100 kilohertz, and the results can be plotted into a SWR curve as shown in figure 8B. If a minimum SWR figure of 1.0/1 is not obtained, or if the resonant frequency of the system is not near the center of the band, a second series of tests should be run.

The transmitter is now tuned to the frequency giving the lowest value of SWR found in the previous test. In the example shown in figure 8B, this frequency is 14,105 kilohertz. The minimum SWR at this frequency is 1.4/1. It is desired to obtain a minimum SWR at a frequency of 14,200 kHz. Now, with the transmitter tuned to 14,105 kHz the gamma capacitor and gamma wire are again readjusted for minimum SWR, which now turns out to be 1.2/1. Varying the transmitter back and forth across the band shows the minimum SWR to be 1.1/1 at 14,160 kHz. The transmitter is now left tuned to the latter frequency and the gamma adjustments are repeated once again. The SWR value drops slightly to 1.08/1, and the new resonant frequency of the system is found to be 14,180 kHz with a SWR reading of less than 1.05/1. Since this is reasonably close to the design frequency and the SWR reading is very low, the tests are concluded.

A Quick and Rapid Test

In actual practice, this series of tests can be telescoped into a single test wherein the capacitor, gamma wire, and transmitter frequency are all varied by small degrees while the SWR meter is observed for minimum reading. An increase in SWR indicated that the adjustment being made at the moment is heading in the wrong direction, and a decrease in SWR means the particular adjustment is in the proper direction. With practice (and an assistant on the ground to vary the frequency of the transmitter) the whole series of adjustments may be made in a matter of a few minutes with a minimum of effort and confusion.

It must be remembered that while a SWR reading of 1.0/1 at the resonant frequency is a comforting state of affairs, the time spent to achieve this must be weighed against the operating advantages gained by a low SWR factor for the antenna system. If the minimum SWR reading turns out to

Base of tower shown on page 48 is hinged to facilitate antenna experiments. Tower base is bolted to a heavy plate hinged to two wood blocks mounted to the wall of the house. Block and tackle permits the tower to be raised easily.

be (for example) 1.3/1 at some frequency within the amateur band it is problematical if the work involved to drop this reading to 1.0/1 is worth the effort. On the other hand, if the system is badly out of adjustment and yields readings of the order of 2.0/1 or so, time taken to readjust the matching system will be decidedly worth while from the standpoints of system efficiency and ease of operation.

A High Impedance Feed System

Quad antennas exhibit terminating impedances as high as 150 ohms or so. A high impedance Quad may be matched with a high impedance coaxial line and a balun. 125 ohm cable exists (RG-79/U and RG-63/U), as well as 150 ohm cable (RG-125/U). Unfortunately, these special cables are not carried in stock in all radio stores, and they are expensive, too, when found. Comparable results may be obtained by the use of a 50 ohm transmission line and a quarter-wave linear transformer made out of 72 ohm line (RG-11/U). This combination will provide a terminal point of about 130 ohms, which can be used with high impedance Quads with good results.

CHAPTER VII

The Tri-Gamma
Multiband Quad Antenna

A multiband Quad is a simple, lightweight, and inexpensive antenna well suited for operation on 14, 21, and 28 MHz. The concentric Quads require no loading coils or traps or other "gadgets" associated with three band parasitic arrays. Best of all a three band Quad has little wind resistance and weighs but a fraction more than a single band antenna.

One of the drawbacks of a multiband Quad has been that three separate feedlines were required, one for each band, unless the driven loops are tied in parallel and fed from a single line as discussed in Chapter VI. This system is simple to build but it requires the use of an antenna tuner and does not permit the impedance adjustments necessary for a low value of SWR on the line.

THE TRI-GAMMA MATCH

The modification and adaptation of the popular gamma match device to the Quad antenna offers a solution to the problem of a common feed system that will function with a multiband antenna of this type. The use of separate gamma matching devices (one for each Quad) allows a relative degree of isolation to be achieved between the antennas while permitting them to be excited from a single transmission line. The best method of interconnecting the gamma devices could only be determined by experiment so considerable time was spent trying various feed systems in order to determine which one provided the greatest degree of flexibility, yet introduced a minimum of undesirable interaction between the antennas.

The first experimental matching system employed three gamma assemblies connected to each other and to the transmission line by short lengths of coaxial cable. It was possible to feed the system at any one of the three gamma devices. This configuration showed promise, but the amount of undesired reactance coupled into the feed point by the presence of the two unused antennas made the tuning procedure of the other antenna a complicated and time consuming task.

Removing the unused antennas did not seem to improve matters as the interconnecting lengths of coaxial line seemed to be causing most of the trouble. Accordingly. the feed system was scrapped and a new one was built using an open wire connecting line. In addition, a reactance capacitor was found to be necessary at the terminating point of the 20 meter loop.

Building the Tri-Gamma Matching System

The assembly of the Tri-Gamma matching system is shown in figure 1. The heart of the device is a short length of open wire transmission line seen running between the center points of the three driven loops of the multi-band Quad. The loops are not broken at the centers of the lower section but are closed, forming a resonant circuit. One wire of the transmission line connects the center points of the loops to each other and to the outer

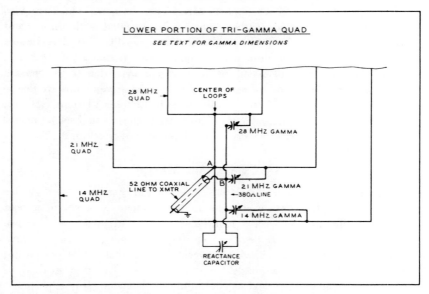

Fig. 1 Tri-Gamma feed system is well suited to 20-15-10 meter Quad. The gamma wires are adjusted to reduce interaction as well as to provide a proper impedance transformation. The gamma capacitors are used to resonate system.

shield of the coaxial transmission line. The other wire of the line connects the terminating points of the three gamma devices together, and to the center conductor of the coaxial line. While the coaxial transmission line may be attached to any point on the open wire line, the least amount of interaction is present when the "coax" is attached to the system at the point of termination of the gamma device of the middle-sized Quad.

The short section of open wire line should not be thought of as an extension of the coaxial line, rather it is part of the Tri-Gamma system. The point of termination of the Tri-Gamma is the junction between the open wire line and the coaxial transmission line.

The individual gamma devices are made of #12 solid copper wire and a small variable capacitor. The wire length and spacing to the Quad loop are set to the preliminary dimensions given in figure 7 of chapter 6. The open wire line may be made of two #12 copper wires spaced $\frac{3}{4}''$ apart. Small spacers made of plastic or other insulating material may be employed to hold the wires parallel to each other at the desired spacing. For low power, open wire "TV ladder-line" may be used.

For powers up to 100 watts or so small receiving type variable capacitors may be used in the gamma assemblies. Higher power levels require capacitors having a greater air gap. The capacitors must be protected from the weather or they will quickly rust and become inoperative. An inexpensive covering may be made from a small plastic refrigerator jar or plastic cup. Seal the lid of the jar with vinyl tape to make it waterproof and coat the gamma wires with roofing compound at the points they leave the enclosure to prevent water from seeping down the wires into the jar. The capacitors are mounted so they can be adjusted before the lid is placed on the jar. A small hole is drilled in the bottom of the jar to allow condensed moisture to escape and the jar is sealed.

Interlocking Effects

The adjustments of the Tri-Gamma tend to be interlocking, as is true of any multiband matching system. If the capacitors and gamma wires are set to the approximate data given in figure 7, chapter 6 before adjustments are made the alignment procedure will be greatly simplified. In general, the gamma capacitor is used to tune the individual Quad system to resonance and the length of the gamma wire determines the impedance transformation required for proper operation. It is to be noted, however, that the use of multiple matching systems introduces unwanted reactances and it will be found that some compensation must be made on each band for the detuning effect of the unused gammas. That is, the 14 and 21 MHz gammas tend to upset the 28 MHz adjustment; the 21 and 28 MHz gammas upset the 14 MHz adjustment; and so on. Fortunately, the reactance capacitor at the 20 meter loop position is used to counteract the effects of detuning introduced by the multiple gamma devices.

The approximate capacitance of the reactance capacitor is 200 pF (200 uufd) and the value is not especially critical. Before tuning begins, a broadcast-type tuning capacitor of 350 pF may be used. Once the correct value of capacitance is determined by adjustment, the capacitor may be removed, measured on a bridge, and a fixed capacitor of the correct value substituted in its place. Individual gamma adjustments will be made easier if the reactance capacitor is set to about 200 pF before tuning begins. By a series of adjustments it is possible, then, to arrive at gamma settings that will deliver a SWR value of 1.0/1 on the transmission line at the resonant frequency of each Quad antenna. If gammas and capacitors are pre-set to the approximate values, the whole tuning procedure should be rapid and painless.

The Test Set-up

In order to evaluate and adjust the Tri-Gamma device it is necessary that the individual gammas be easily reached and that the test instruments be mounted near the center of the Quad array. At the same time the array should be reasonably in the clear and high enough so that ground effects are at a minimum. In addition, if it is possible to raise the antenna to its final height during and after the tests it is simple to determine the results of small changes made to the system.

The following adjustments were performed on a three band Quad mounted on a fifty-five foot heavy duty "crank-up" tower. When the tower was retracted the Quad rested with the center of the assembly about twenty feet above ground. Fully extended, the tower raised the Quad completely clear of surrounding objects. By climbing atop the roof of the house and standing on a small platform it was possible to make adjustments to the matching devices mounted on the antenna. This set-up proved satisfactory in every respect for the investigation, and little difference in measurements was noted when the array was raised to the full height of the tower.

ADJUSTING THE TRI-GAMMA MATCH

A suggested test set-up for Tri-Gamma adjustment is shown in figure 2. A low power exciter and a SWR meter in the transmission line to the Quad are required. Alternatively, a dip-oscillator and an Antennascope may be used for adjustment. Full details for construction of an Antennascope are given in the *Radio Handbook*, published by *Editors And Engineers* division, Howard W. Sams & Co., Indianapolis, IN 46268. It is recommended, however, that the set-up in figure 1 be followed as it provides a quick and easy means of determining transmission line SWR.

Fig. 2 SWR meter and exciter are used to adjust Tri-Gamma match. Reactance capacitor is pre-set, then gamma capacitors and rod lengths are varied for lowest SWR. "Touch-up" is done with reactance capacitor. Reactance matching system was devised by W6CHE.

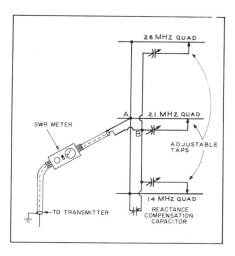

Adjustments for the Tri-Gamma Quad are carried out as follows:

1—The Tri-Gamma system is pre-set to the approximate dimensions and values given in figure 1 and in figure 7, Chapter VI. The feedline is attached to the matching system, with a SWR meter placed in the line near the antenna where it may be easily observed, as shown in figure 2. Make sure the shield of the coaxial line goes to *point A* of the Tri-Gamma system and the inner conductor is connected to *point B*.

2—A small amount of power is applied to the system from the exciter at 10 meters. The 10 meter gamma capacitor and gamma length are adjusted for minimum SWR indication. The exciter is moved about in frequency until the lowest SWR is indicated. then the gamma capacitor and length are again readjusted until the SWR is at the lowest possible value.

3—The exciter is switched to 15 meters and the 15 meter matching section is adjusted for minimum SWR on this band, in the manner described above.

4—The exciter is switched to 20 meters and the 20 meter matching section is adjusted for minimum SWR on this band. The reactance capacitor is now adjusted to enhance the SWR null.

5—The 15 and 10 meter bands are now rechecked for minimum SWR, which may have risen after adjustment of the reactance capacitor.

It will be noticed that the 20 meter gamma section has the greatest detuning effect upon the assembly. Exact adjustment of the 10 meter and 15 meter gammas may be carried out with little interaction as long as the 20 meter gamma capacitor is set a minimum capacitance. As soon as the 20 meter gamma capacitor is brought into play, however, an intolerable detuning action is noticed on the 10 and 15 meter sections unless the reactance capacitor is used to compensate for the ill effects of the 20 meter gamma section. The experimenter can easily tell if the adjustments are getting out of line, as it seems that the SWR improves as the gamma lengths are shortened. If the gamma lengths are much less than noted in figure 7, Chapter VI, it is a good indication that the reactance capacitor setting is incorrect.

Once the adjustments have been completed, the SWR response of the array should resemble the curves shown in figure 4, Chapter IX. It may be noticed that the SWR will tend to rise more rapidly on the high frequency side of the resonant frequency, as compared to the rise on the low frequency side of resonance. This is normal and is due to the fact that the frequency of operation is approaching the self-resonant frequency of the director elements.

The experiments conducted in this chapter were performed, in part, by W6CHE on his four element, tri-band Quad antenna shown in Chapter X. Our thanks are extended to W6CHE for permission to describe the unique reactance tuning system perfected on the antenna illustrated.

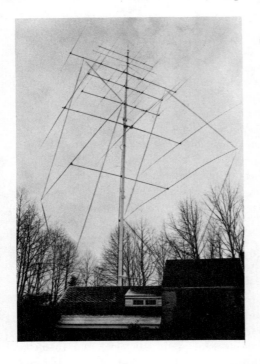

80 meter Quad dwarfs 115 foot "Christmas Tree" antenna system of Joe Hertzberg (ex-K3JH). Two diamond-shaped loops were used, about 50 feet on a side, plus 18-foot adjustable stubs on each element. Element spacing was 36 feet. Old Timers remember impressive signal this giant beam had on 80 meters.

CHAPTER VIII

Build Your Own Quad Antenna

The use of light, thin wire elements in lieu of heavy aluminum tubing greatly simplifies the task of building a beam antenna. The weight of a full size two or three element parasitic array is approximately 60 pounds, of which 25 pounds is the weight of the elements and 35 pounds is the weight of the boom, supports, and brackets. The 20 meter Quad on the other hand requires less than three pounds of copper wire elements and the weight of the supporting framework may be held to less than 25 pounds. The Quad therefore enjoys a two to one weight advantage over a comparable parasitic array. The wind resistance of the Quad antenna is also measureably lower than that of the parasitic array.

With the addition of extra wire loops on the framework the Quad is easily converted to a three band antenna, exhibiting little extra weight or wind resistance over the single band Quad, and requiring no trap circuits or loading coils of questionable efficiency. The Quad is truly a remarkable antenna for the weight, cost of materials, and assembly time!

THE WOOD AND BAMBOO FRAME

The simplest and least expensive Quad assembly is constructed of bamboo "arms" and a wooden supporting structure as shown in Figure 1. Four bamboo poles are required for each Quad loop. bolted to a wooden center plate by means of galvanized U-bolts. The center plates in turn are bolted to opposite ends of a wooden beam that is attached to the support and rotating structure.

The bamboo poles chosen for the element arms should be clean, straight, and free of splits and cracks between the rings. Extra-length poles should be

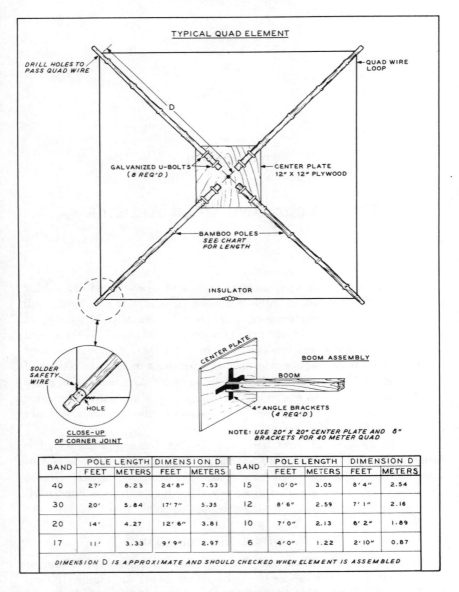

BAND	POLE LENGTH		DIMENSION D		BAND	POLE LENGTH		DIMENSION D	
	FEET	METERS	FEET	METERS		FEET	METERS	FEET	METERS
40	27'	8.23	24' 8"	7.53	15	10' 0"	3.05	8' 4"	2.54
30	20'	5.84	17' 7"	5.35	12	8' 6"	2.59	7' 1"	2.16
20	14'	4.27	12' 6"	3.81	10	7' 0"	2.13	6' 2"	1.89
17	11'	3.33	9' 9"	2.97	6	4' 0"	1.22	2' 10"	0.87

DIMENSION D IS APPROXIMATE AND SHOULD CHECKED WHEN ELEMENT IS ASSEMBLED

Fig. 1 Inexpensive Quad assembly is made of bamboo elements, wood center plate and 2" x 2" wood boom. Bamboo poles are wrapped with vinyl tape between the joints to strengthen them and are given two coats of shellac for protection against the weather. Wood center plate is given two good coats of house paint to seal edges of plywood. All hardware is galvanized to minimize rust and corrosion. Antenna should be laid out on ground and wire stretched around the framework before the wire holes "D" are drilled in the bamboo poles.

Fig. 2 Commercial aluminum spider is designed for use with bamboo or fibreglass arms. Aluminum boom is used in this antenna. The ends of the boom are plugged with wooden inserts for added strength. Arms are locked to spider with hose clamps.

purchased so that the small tips may be cut off and discarded. The poles can be purchased at bamboo distributors in large cities, at some rug stores in similar towns, or perhaps at garden nurseries or through large mail order houses. The poles should be wrapped firmly with electrical vinyl tape between the joints to retard splitting and are then given two coats of outdoor varnish or shellac to protect them from the weather.

Plywood is ideal material to use for the center plates of the Quad framework. The plates measure about one foot long on a side and are cut from 5/8-inch stock. It is necessary to seal the edges of the plywood against moisture to prevent the plate from cracking or splitting. Two liberal coats of good outdoor house paint will do the job. The plates are drilled to pass U-bolts which clamp the bamboo poles along the diagonal axis of the plate as shown in the drawing. Plated or galvanized U-bolts, washers, and nuts are used in the assembly to retard rust and corrosion. The butt ends of the poles are wrapped with vinyl tape for added strength at the points the U-bolts contact the bamboo. Two bolts are required for each pole and the poles are positioned so that there is a gap of about $1\frac{1}{2}$-inches between the butt ends. Washers are placed under all nuts to prevent them from digging into the soft surface of the plywood. Be certain all hardware is rust-proof, or you will have an unhappy time when you attempt any repair work, or when you try to dismantle the beam.

The boom should be a section of dry 2″x 2″ lumber, well painted to protect it from moisture in the air. "Green" lumber will tend to warp as it gradually dries out, imparting a nasty twist to the symmetrical Quad design! Sanding the boom before painting it is a wise measure as this action will protect you from slivers and splinters during the assembly process.

The center plates are attached to the ends of the wooden boom by means of four plated steel angle brackets as shown in the drawing. The brackets are mounted slightly off-center on the boom so that the retaining bolts will not interfere with each other passing through the boom. Satisfactory brackets can usually be found in a hardware store.

Waterproof Your Bamboo Arms with Fibreglass

The bamboo arms, even when coated with shellac and wrapped with vinyl tape, have a short life. Wind, rain, and sun tend to shrivel and split the soft bamboo, and sooner or later the arms will warp out of shape or permit the Quad wires to sag.

You can give the bamboo a protective coat of *Fibreglass* compound that will ruggedize them and extend their life indefinitely. The treatment costs very little and takes only a few minutes. This is how you do it:

Materials for Fibreglass treatment can be obtained at Marine hardware stores, large building supply establishments, and some mail order houses. You will need a roll of three inch Fibreglass cloth, and a can of Fibreglass liquid cement. One name for this liquid is *"Boat Resin."* When you buy the liquid you will also get a smaller can of solvent to mix with the resin.

The first step is to lay the bamboo poles out between two boxes or other supports so that the bamboo is clear of the ground. Clean the poles with a damp cloth to remove dust and dirt, and spirally wrap each pole with the Fibreglass cloth until it is completely covered. You can hold the cloth in place with rubber bands or narrow strips of paper "masking tape." The cloth windings should overlap about an inch as you wind the material about the pole. The last step is to mix the solvent with the resin according to the instructions, and give each cloth covered pole a liberal coating of the resin. Use an old paint brush and work smoothly and rapidly from one end of the pole to the other. Let the poles dry over night, then give them a second coat of resin. When the liquid dries, it forms a firm, hard intermixture with the Fibreglass cloth. The treated bamboo pole is now exceedingly strong and you need never worry about the pole cracking or splitting. Caution: do not breathe the fumes of the resin!

Assembling the Antenna

You will find that the wood and bamboo framework is a flimsy and unwieldy structure, having as much structural strength as a jellyfish. However, once the antenna wires are strung in position the assembly will magically become neat and strong and amazingly rigid. Your next job is to string the antenna wires on the framework. Remove the end framework from the boom during the following steps and lay them on the ground.

Since the Quad loops are of a predetermined size you cannot take up slack in the wires by tightening the loops. Rather, the slack in the wires (if any) must be absorbed by expanding the bamboo framework until the wires are under tension. Drilling data for the holes in the bamboo poles to accommodate the antenna wires is given in figure 1. Final wire tension may be adjusted by spreading the poles apart at the center plate before the U-bolts are tightened. Begin by cutting the Quad wires to the dimensions given in figure 16 chapter 3, or figure 5 chapter 5. Allow enough extra wire on each loop to make the end connections. If you are making a 3-band Quad construct the outside (largest) loop first. Clean the ends of the wire and thread it through the holes drilled in the tips of the bamboo poles. The ends of the wire should meet at the center of the bottom side of the loop. Temporarily attach the wires to the center insulator so that you can adjust the tension, making sure that the insulator remains in the center of the side of the square. Remeasure the sides of the loop. When everything is "ship-shape" wire each bamboo pole to the loop. Scrape the enamel covering from the Quad wire for an inch on each side of the poles and pass a short wire jumper over the pole, wrap it around and solder to the antenna wire on each side of the pole. This safety wire will prevent the loop from shifting about on the framework.

You can now string the remaining wires for the other bands on the framework. It is not necessary to use jumpers on these loops The configuration of the assembly, however, has been fixed by the large loop and you cannot adjust the tension of the inner wires by adjusting the poles within the U-bolt clamps. Tension is not particularly critical on these wires and if they are cut to the proper dimensions they will fall into place. If adjustment is desired it is permissible to drill a new mounting hole in one arm above or below the old hole and adjust the tension by passing the loop wire through the new hole. Make sure that the three center insulators fall one above the other and that the loops are completely symmetrical.

The second loop assembly may be made by laying the components atop the first one and making a "Chinese copy". When it is completed, reflector stubs should be attached to one of the loop assemblies. The last step is to solder all joints using a *hot* iron and rosin core solder. A bad joint in the assembly would certainly cause havoc to proper beam operation.

The final operation is to mount the end frameworks to the center boom. A little thought should be given to this operation because once the Quad is assembled it becomes an unwieldy object. A 28 MHz Quad may be handled with ease by one person. It takes two to manipulate a 21 MHz Quad, and a 14 MHz Quad is quite a handful for three men. In this respect the Quad resembles a porcupine: there is no "handle" to grab it. You cannot lift it by the bamboo poles as this will tend to warp the assembly, and the boom of

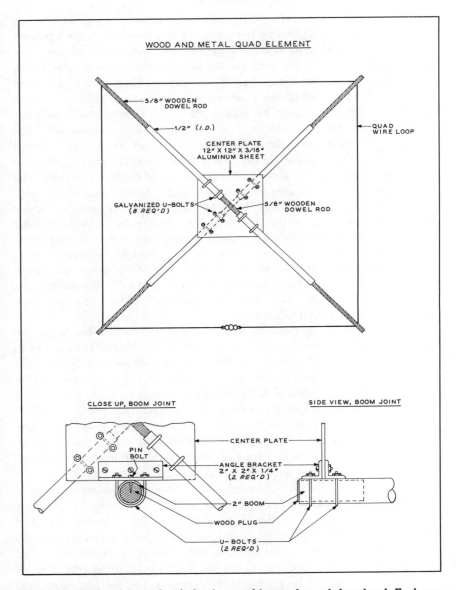

Fig. 3 Sturdy Quad is made of aluminum tubing and wood dowel rod. Each arm of array is made of sections of tubing and rod. Arms are then attached to a 12" x 12" aluminum center plate with U-bolts. One arm is placed on each side of the plate. The assembly is held to the 2" aluminum boom by means of double angle brackets and U-bolts. Assembly is pinned to the boom by bolts running through the bracket, boom and wood plug.

the 14 MHz Quad is too high in the air to grasp easily. The Quad should therefore be assembled in a clear space near your tower, and a short section of wood or pipe can be mounted vertically beneath the center boom to facilitate moving the Quad about.

THE ALUMINUM FRAME

While the bamboo and wood framework provides a satisfactory assembly a stronger, more rugged, and longer lasting one may be built of metal. Alignment of the metal framework is permanent and there is no danger of cracking or splitting poles which is always possible with the wooden framework. A suitable metal Quad is shown in figure 3. The assembly comprises a boom made of a section of two inch aluminum electrical conduit or irrigation pipe, center plates made of 3/16" sheet aluminum, and arms made of aluminum tubing or electrician's EMT steel tubing, with wooden dowel extension tips.

All parts of this metal framework are at ground potential with respect to the antenna and care must be taken to make sure the metal arms are not self-resonant at the operating frequency of the array. To achieve this, the arms must not come within close proximity to the Quad wires. It is necessary therefore to employ insulated tips at the extremities of the arms to act as insulators and separators. Inexpensive wood dowel rod (well varnished) may be used. Lengths of phenolic tubing will work as well.

It is possible to obtain electrical metal conduit (EMT) for the diagonal arms. EMT tubing is available in 10 foot lengths, so two pieces spliced at the center are required for a 20 meter Quad. Dowel tubing, $\frac{5}{8}$-inch in diameter will just fit within the so-called "$\frac{1}{2}$-inch" diameter EMT tubing, which actually is somewhat over $\frac{5}{8}$-inch inside diameter. Arm assembly is shown in figure 3.

The completed assemblies are bolted to opposite sides of the center plate as shown in the drawing. The dowel extensions are drilled for the Quad wires, and the loop is installed and safety-wired into position at each pole as previously described in this chapter. The next job is to attach the frameworks to the boom. The simplest and best arrangement is to clamp the center plate to the boom with the aid of an aluminum bracket and large U-bolts. To prevent the clamps from crushing the boom it is necessary to plug the end of the boom with a block of wood. A piece of dry 2"x 4" lumber about six inches long shaved to a circular contour will do nicely. Make two plugs, one for each end of the boom. The center plate is bolted to two 12" lengths of 2"x 2"x $\frac{1}{4}$" aluminum angle stock placed back to back on each side of the plate. The angles are firmly fastened to the boom by two U-bolts. One U-bolt is on each side of the center plate, imparting longitudinal rigidity to the framework. The U-Bolts by themselves are not sufficient to prevent the assemblies from being rotated about the boom under heavy gusts of wind. It is therefore necessary to drill the angle plate, boom, and wooden plug to pass a ¼-inch bolt which acts as a pin holding everything in position.

The Mini-Quad "Spider" Support

A two element Quad having optimum element spacing on each band may be built on a "Spider" structure such as shown in figure 4. Built of iron pipe, this simple, welded framework will accommodate bamboo or fibreglas arms of proper length for a 20, 15 or 10 meter Quad, or an interlaced tri-band version. The "spider" is made in two parts so that the elements may be assembled on the ground and carried to the top of the tower for final assembly. Boom length is only two feet so the entire antenna can be easily carried by a single man.

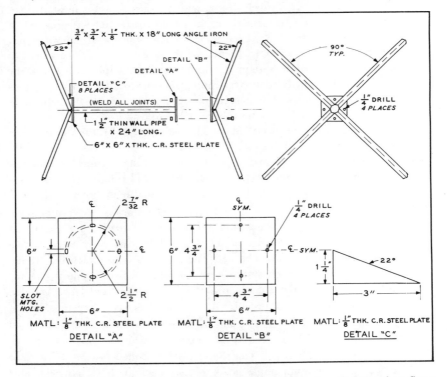

Fig. 4　Metal support for two element Quad reduces boom length to two feet. Structure is made in two parts so that Quad elements may be assembled on the ground and carried to top of tower for final assembly. Made of welded pipe, the Quad "spider" should be given two coats of paint before use.

CHAPTER IX

Quad Tuning and Adjustment

Dimensional data given in this Handbook will enable the Quad builder to assemble and install his antenna with the assurance that it is tuned and adjusted for near-optimum results when located well in the clear at an elevation of forty feet or better. The only tuning adjustment that can be made once the dimensions of the antenna are fixed is placement of the reflector stub, or adjustment of the director element (if used). Stub adjustment does little to vary the forward gain of the array, a mis-adjustment of several inches on the stub shorting bar will lower the gain of the array less than 0.1 decibel. On the other hand, the front to back ratio of the array is quite critical as to stub placement.

The F/B ratio is a function of stub adjustment, as well as of antenna placement and character and proximity of nearby objects. This is the reason why two identical Quad antennas will exhibit different F/B ratios in different locations and at different heights above the ground. The F/B ratio is also affected by the angle of arrival above the horizon of the incoming signal. High angle signals arriving at the rear of the array, or vertically polarized signals produce less F/B ratio than do low angle horizontally polarized signals.

If optimum front to back ratio is desired it is necessary to adjust the stub of the parasitic element after the Quad antenna has been placed in the *approximate* position of operation. If you can climb your tower and adjust the reflector stub while the antenna is in the operating position, well and good. You can achieve optimum results. In many cases it is impossible or

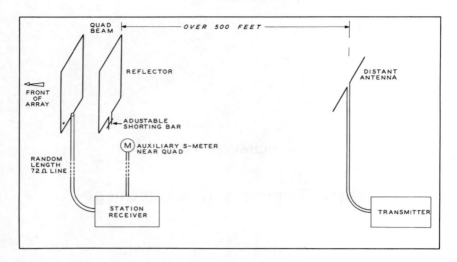

Fig. 1 Quad antenna may be adjusted for maximum front-to-back ratio with this test set-up, making use of your receiver and the transmitter of nearby ham. The distant test antenna should be horizontal for optimum results.

risky to attempt this tuning task, so the next best procedure must be used. Very good results can be achieved with the Quad positioned with the lower wires resting about fifteen or twenty feet above the ground. If you use a telescoping tower, the adjustments may be made with the tower retracted, or the Quad can be temporarily placed on a short wooden mast mounted atop a garage or building. Raising the Quad into the final operating position will alter the tuning adjustment somewhat, but the deterioration in the F/B ratio will not be excessive.

Antenna Adjustment

A typical test set-up for measuring F/B ratio on a Quad antenna is shown in figure 1. A shorting bar made of two alligator clips mounted back to back is placed on the reflector stub wires and the driven element of the Quad is attached to the station receiver by means of a length of 72 ohm "TV-type" twin lead. The S-meter of the receiver is temporarily disconnected and a substitute meter is placed at the adjustment position connected to the receiver by a length of two wire lamp cord. The signal strength reading of any signal tuned in on the receiver can thus be measured while adjustments are made to the reflector stub.

The assistance of a nearby amateur is now required to supply a steady signal for adjustment purposes. The antenna used by your friend should be

Three band Quad of DX er W9FKC has proven ability in DX contests! Quad is mounted atop wooden tower. 21 and 28 MHz interlaced sections allow operation on three bands. Main band of operation is 14 MHz.

horizontally polarized—using a signal source having a vertical radiated field pattern will lead to confusing results. The back of the array is now aimed at the signal source. Receiver gain is adjusted to produce a S-9 meter reading and the shorting bar is then moved up and down the stub until minimum S-meter reading is obtained. The point of minimum response may be very broad or very sharp, depending upon the mode of arrival of the test signal. If several nearby amateurs can be urged into serving as signal sources, it might be found that a different stub adjustment is required for each source. Multi-path propagation and signal reflection from nearby objects is responsible for this seemingly confused situation. An average of the various stub settings can be chosen, or a signal several thousand miles away can be used for a final check. If your receiver is tuned to the edge of the phone band (and you have earphones with a long extension cord) you can easily null the reflector stub on a distant signal for optimum F/B ratio.

Antenna Installation

You should mount the Quad antenna at least thirty feet in the air (as measured to the bottom wire) for best results. Unlike other antennas, the Quad will still provide a reasonably low angle of radiation at lower elevations, but every effort should be made to provide the best possible antenna site. Optimum results will be obtained when the antenna is forty to fifty

The 14 MHz single band Quad antenna mounted atop a flat roof for experiments.

feet above the ground and there are no nearby telephone or utility lines. The heavy duty "TV-type" crank-up towers are relatively inexpensive and rugged and are recommended for use with the Quad antenna. When guy wires are used to steady the antenna tower they should be broken into six foot sections by strain insulators to prevent resonance effects in the wires that might interfere with proper antenna operation.

ANTENNA EVALUATION

"How does it perform?" That is the universal question to be asked once the antenna is placed in operating position and the first few contacts are made. The haphazard method of comparing signal reports with nearby amateurs at a distant point may provide some comfort to the beam owner, but will provide little in the way of knowledge as to the efficiency of his antenna system as a whole. However, if the antenna user can "hold his own" against similar arrays in the neighborhood driven by transmitters of like power he can at least be reassured that some of the r-f supplied to the antenna is doing a bit of good!

A standing wave ratio meter placed in the transmission line will give a good picture of general antenna operation. The SWR readings should be

The four element Quad is ready to go up in the air. This 15 meter array employs a 22-foot boom with top and side guys. Power gain of the antenna is better than 9.4 decibels, with a front-to-back ratio of about 20 decibels.

taken at 100 kilohertz points across the band of operation, and a curve plotted, showing the SWR readings as a function of the operating frequency. Typical SWR curves for 10, 15, and 20 meter two element Quad antennas are shown in figure 4. The curves bear the same general shape as the SWR curves for the two and three element parasitic arrays but have greater bandwidth (as defined by the points of 1.75/1 SWR). In the case of the 20, 15, and 10 meter Quads, the operating bandwidth is appreciably greater than the width of the amateur band. The frequency of resonance of each antenna is indicated on the SWR graph, and is controlled by the dimensions of the Quad loop. In each case the resonant frequency falls near the center of the amateur band. Making the Quad loops slightly larger will lower the resonant frequency, conversely making the loops a bit smaller will raise the resonant frequency.

If a perfect impedance match is achieved between the antenna and the transmission line a SWR of 1.0 to 1 will be achieved at the resonant frequency. Unless a variable impedance matching device (such as the Gamma match) is employed the minimum SWR usually will not drop to 1.0 unless a lucky combination of circumstances are encountered. Size of wire, slight physical differences in construction, and variations in transmission line impedance between various manufacturers will all combine to make the perfect theoretical match somewhat less than perfect. Resonant SWR readings of up to 1.5/1 will usually be encountered and are perfectly acceptable. The only penalty that must be paid for a high value of SWR on the transmission line are the even higher SWR values that must be accepted at the edges of the band. In some instances odd lengths of transmission line must be employed to permit the pi-network transmitter to load into a coaxial line having a relatively high value of SWR.

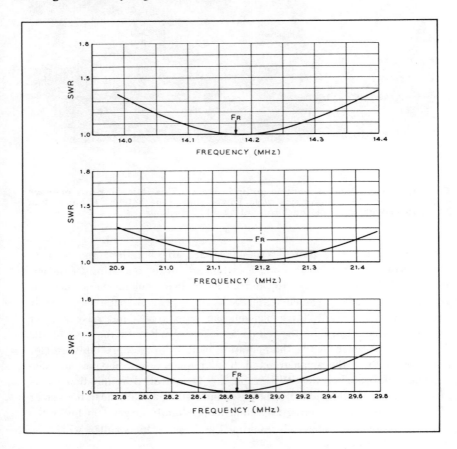

Fig. 4 Properly adjusted Quad antennas show 1.0/1 SWR at resonant frequency of driven element, with SWR curve rising smoothly and gradually each side of resonance point. Bandwidth of Quad is ample for complete coverage of 14, 21 or 28 MHz band. Three band Quad exhibits similar curves. Use of gamma match with Quad antenna permits quick and easy adjustments.

Three element Quad antenna. The boom is supported at ends by a top guy. Antenna gives performance comparable to four element Yagi beam at fraction of the cost of the all-metal antenna.

Antenna Maintenance

As a droll wit once said, "Getting the antenna up in the air is easy! Getting it *down* is the hard job!" This is true of any antenna, and the Quad is no exception. Moisture in the atmosphere and the effects of sun, wind, and rain are at work to corrode metal and rot wood. An antenna exposed to the elements will quickly deteriorate unless important steps are taken to protect the array against the weather. Follow these simple, effective rules and you will have little trouble with your Quad. When the day comes to take it down, you can accomplish this feat with a minimum of trouble!

1—Never, *never* use unplated hardware at any point in your antenna system! Use stainless steel. Next best is cadmium or galvanized plated steel, although this will rust eventually. Unplated parts are worthless and almost impossible to remove, once rusted.

2—Paint all exposed wood surfaces with a primer coat followed by two coats of high quality oil-based outside paint (no latex or water-based paint!) Coats should not be too heavy as this won't last — they will peel off. A dull, blue color is suggested to blend with the sky.

3—Aluminum components (the boom and end plates of the all-metal Quad, for example) should be thoroughly cleaned with sand paper and given a thin coat of *Zinc Chromate* primer. This green metal priming paint is now available in *Aerosol* spray dispensers for easy application. *Caution!* Do not breathe the Zinc Chromate fumes. They are toxic and prolonged inhalation will make you ill. An antidote to the fumes is a glass of milk.

When the primer coat is dry, the aluminum may be given a coat of aluminum paint, or dull blue housepaint.

4—Coaxial lines (RG-8/U and RG-57/U, for example) should be sealed against excessive moisture. The black vinyl jacket of the line is relatively waterproof, but moisture will seep into the line at the end joints

Experimental three band concentric Quad uses close-spaced reflector for 21 and 28 MHz and wide-spaced reflector for 14 MHz. Average impedance of all three beams is close to 75 ohms for a good match to RG-11/U coaxial feedline.

Important! Keep your antenna clear of all power and telephone lines. Amateurs have been electrocuted in the past when erecting their antenna.

where the jacket has been cut. Seal each joint and coaxial plug with a wrapping of vinyl tape and your line will last a long time.

5—An ounce of prevention is worth a pound of cure. Keep an observant eye on your antenna. When it shows sign of weathering, take it down for an overhaul job. Make sure your guy wires are in good shape and replace them if they start to show rust marks.

Final Assembly of the Quad Antenna

You will find that a 3-band Quad or 14 MHz Quad is a light but bulky affair that has a tendency to tangle with the guy wires of the tower. If you have a guy-less tower or a telephone pole you are in luck, as you can hoist the complete Quad to the tower top with little effort. It is often easier to erect the Quad in three separate sections; the boom first, then one Quad element, followed by the other. If you do this it is mandatory that the boom of the antenna be capable of sliding back and forth in its retaining mount atop the tower in order for the assembler to be able to reach the ends of the boom. A good safety belt is a necessary piece of equipment for this undertaking. If the Quad has been preassembled on the ground the man atop the tower can be reasonably sure the affair will go together when it is in the air. The unfortunate assembler at the top of the tower is usually at a disadvantage in that he requires one hand to hang onto the tower! Assembling the Quad framework with one hand is not easy, but it can be done if the parts are hoisted to the top of the tower by a ground-based assistant. A stout rope tied about the center plate of the Quad will prevent it from getting away from the control of the assembler.

CHAPTER X

The Quad Round-Up

The versatility of the Quad loop has provided many interesting innovations in the basic design. Some of the more popular variations are discussed in this chapter, along with information on 4- and 5-element "Monster Quads" for 20, 15 and 10 meters. Finally, an evaluation between the Quad and the Yagi antenna is given.

THE QUAD ELEMENT

No law restricts the Quad element to a square or a diamond shape. While these forms are the easiest to assemble, Quad antennas have been built with circular elements and with triangular shaped elements. The high efficiency of these various shapes leads to the conclusion that the physical arrangement of the Quad element is relatively unimportant, the gain being highest when the wire element extends around the greatest area for a given element length. Since the circle is the configuration which meets this requirement, it may well be that a circular Quad would have slightly higher gain than any other shape, but the difficulty of building such elements has prevented any meaningful data from being accumulated.

The equilateral, triangular-shaped Quad element shows promise, as it may be arranged for mounting from an apex point in a simple fashion, or may be used as a fixed inverted element slung between two supports, as suggested in Figure 1. Two or more triangular Quad elements may be combined to form a *Delta Quad* beam antenna.

The Quad element by itself, moreover, may be used as a simple beam antenna, having about 1.4 decibel gain over a dipole with a figure-8 radiation pattern at right angles to the plane of the element. It can be supported from a single pole and is a recommended antenna where space is restricted and good radiation efficiency is desired.

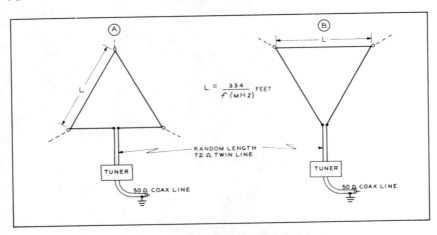

Fig. 1 The Quad element may be arranged in triangular fashion to form the Delta Quad. The antenna element may be suspended from its apex (A) or supported from two poles (B). Either shape may be used in a beam configuration supported from a center boom. Fed as shown, polarization is horizontal. Power gain of the triangular Quad loop is about 1.4 decibel over a dipole. Radiation is at right angles to the loop (into and out of the page). Impedance of single loop is about 120 ohms. Impedance of 2-element Delta Quad is about 80 ohms.

THE EXPANDED QUAD (X-Q) BEAM

A single Expanded Quad element (often referred to as a *Bi-Square* array) provides a bidirectional, figure-8 pattern having a worthwhile power gain of about 5 decibels (Figure 2). As separate wire arrays of this type can be suspended at right angles from a single pole without interaction between them, this offers a solution to the problem of erecting beam antennas in a restricted space.

The Bi-Square beam is fed with a quarter-wave transformer section, a balun and a 50 ohm coaxial line. The impedance of the transformer section is varied by changing the spacing between the conductors, or the conductor diameter, until a good match is achieved to the transmission line.

The shape of the Bi-Square may be altered to fit the available space. It may be rectangular, triangular, square, diamond or circular as the occasion demands. If tuned feeders and an antenna tuner are used with this array, it may be operated at half the design frequency. A 10 meter Bi-Square, in fact, will provide good results on the 15 and 20 meter hands, as well as on 6 meters when used with a tuner.

THE TRIANGULAR LOOP (DELTA QUAD) BEAM ANTENNA

As discussed earlier in Chapter II, the Quad antenna originated from

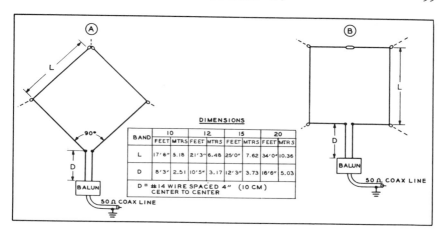

DIMENSIONS

BAND	10		12		15		20	
	FEET	MTRS	FEET	MTRS	FEET	MTRS	FEET	MTRS
L	17'6"	5.18	21'3"	6.48	25'0"	7.62	34'0"	10.36
D	8'3"	2.51	10'5"	3.17	12'3"	3.73	16'6"	5.03

D = #14 WIRE SPACED 4" (10 CM)
CENTER TO CENTER

Fig. 2 The expanded Quad (Bi-Square) element provides a power gain of about 5 decibels over a dipole and makes a good bidirectional beam in itself. Two arrays (for the same or different bands) may be mounted on a single support and switched to provide near-complete coverage. Array is horizontally polarized. Additional information on the X-Q array is given in Chapter V. Two element array will provide nearly 9.5 decibels gain over a dipole, and 3-element array will provide about 12 decibels gain. Ten meter X-Q beam is no larger in "wing-spread" than 20 meter normal-sized Quad antenna. X-Q array will work on harmonic and half-frequency if tuned feeder system is employed.

the idea of a "pulled open" folded dipole. Early Quad configurations made use of a diamond or square loop. Recent tests have been conducted with a triangular shaped loop having a circumference of one wavelength. This antenna shape has often been called the *Delta Quad*. It may be used with the apex of the triangle oriented either down or up (Figure 1). The triangular shaped element seems to exhibit the same amount of gain over a dipole as does the square or diamond shape, and the radiation resistance is about the same as the other versions.

The Delta Quad elements may be mounted above the antenna boom, thus providing a bit more effective height to the array for a given tower height. The sides of the delta may be made of aluminum tubing, leading to a very rugged installation that can withstand high winds and bad weather. An illustration of a typical Delta Quad is shown in Figure 3.

The angle at the base of the Delta Quad is about 75°. The lateral wing elements are made of aluminum and are attached directly to the metal boom, thus providing good lightning protection. The top portion of the loop can be made of wire and adjusted so as to place tension on the wing sections. The wings are about 1/3 wavelength long each and the top wire is slightly shorter. For this model of the Delta Quad, element spacing is about 0.15

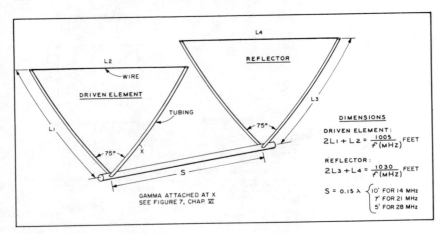

Fig. 3 Delta Quad is formed of triangular elements inverted and supported at the apex. The wings may be made of aluminum tubing and are about 1/3 wavelength long. The top portion of the loop, or triangle, is made of wire. Base angle of the delta element is 75 degrees and both wings may be grounded to the metal boom. Off-center, gamma match feed is suggested, using the gamma dimensions and spacings given in Fig. 7, Chapter VI as a starting point.

wavelength, and gain is estimated to be nearly 7 decibels. Feedpoint impedance (if the driven element is split) is about 70 ohms.

The gamma match described in Chapter VI is also suggested for use with the Delta Quad beam. This system is easy to adjust and permits both halves of the driven element to be mounted securely to the boom, without the need of insulating the wing element from the boom.

Building the Delta Quad

The Delta Quad may be constructed in the usual manner, utilizing a center boom with three aluminum arms, or it may be made of tubing and wire supported from a boom at the apex, as with the design described in this section. Overall dimensions for the Delta Quad are given in Figure 3.

The angle between the wings of the Delta Quad is 75° and the boom must be drilled properly to maintain the same angle in the reflector element as in the driven element. A template or jig may be made of plywood to form a collar fit with the boom, including the 75° angle, as shown in the illustration. A center line is carefully marked along the boom and the template used to align the element holes. A chassis punch of the proper size is suggested for cutting clean holes in the boom. A set of holes for one pair of driven element and reflector wings is drilled and the elements temporarily inserted in the holes. The template may then be used to check the remaining two holes before they are drilled.

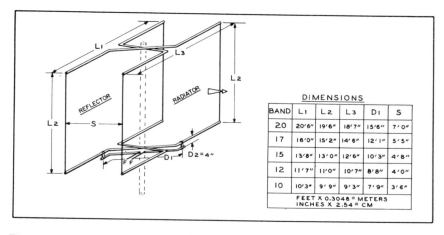

DIMENSIONS

BAND	L₁	L₂	L₃	D₁	S
20	20'6"	19'6"	18'7"	15'6"	7'0"
17	16'0"	15'2"	14'6"	12'1"	5'5"
15	13'8"	13'0"	12'6"	10'3"	4'8"
12	11'7"	11'0"	10'7"	8'8"	4'0"
10	10'3"	9'9"	9'3"	7'9"	3'6"

FEET X 0.3048 = METERS
INCHES X 2.54 = CM

Fig. 4 The Swiss Quad is an all-driven array having a unidirectional field pattern similar to the regular Quad. The elements may be grounded to the supporting mast at the top and bottom. A T-match feed system is used. The Swiss Quad may be fed with an open wire (TV-ribbon) line and antenna tuner or a balun and coaxial line. T-match is adjusted for lowest SWR.

The element wings are passed through the boom and locked in position. The tops of each element are drilled to take a 10-32 rust-proof bolt and the top wires are positioned before the gamma rod is installed. The gamma rod is made of a section of 3/8-inch diameter aluminum tubing flattened at one end and drilled for 10-32 hardware. The flattened end is bent at right angles to fit over the connecting bolt on the gamma capacitor housing, which is made from a plastic box.

THE SWISS QUAD

First used in Europe, the unusual *Swiss Quad* is illustrated in Figure 4. The horizontal elements of the Swiss Quad are made of aluminum tubing and the vertical elements of wire. Both loops are fed by means of a T-match (double gamma match) and the antenna bears a physical resemblance to two stacked W8JK beams, connected together at their element tips.

While no gain measurements are available, on-the-air tests seem to indicate that the Swiss Quad has a gain figure comparable to that of the more common 2-element Quad using a parasitic reflector. The pattern of the Swiss Quad exhibits a deep null off the back, common to arrays of this type where all elements are driven, directivity being obtained by the use of a smaller-than-resonant director element.

The crossover points of each element are grounded to the supporting structure, and it is suggested that the balanced T-match be fed with a balun

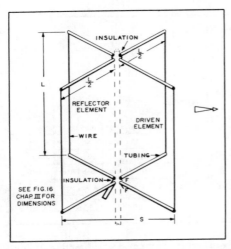

Fig. 5 The G4ZU Birdcage Quad uses two Quad loops folded back at the center so that the top sections act as supports for the vertical wires. Reflector and driven element loops are individually insulated from the supporting mast. The reflector loop uses a stub, or may be made oversize. Driven element is fed from a balun and coaxial transmission line.

from an unbalanced coaxial transmission line. Length and spacing of the T-match is adjusted for lowest SWR on the coaxial line. While the original Swiss Quad design did not show resonating capacitors in the T-match bars, it is suggested that they be included. Capacitance values and rod dimensions shown in Figure 7, Chapter VI, are recommended as a starting point.

THE BIRDCAGE QUAD

An early variation of the Quad design is the *Birdcage Quad* of G4ZU in England. This compact antenna is electrically equivalent to a 2-element Quad having a simplified assembly that solves some of the difficult mechanical problems associated with the standard Quad configuration. The Birdcage uses two full-size Quad loops, the upper and lower portions made of aluminum tubing formed into a supporting cross (Figure 5). These sections are the supporting frame for the vertical portions of the loops, which are made of wire. The Birdcage may be slung from a single pole, without the need of a special supporting structure.

No gain figure is available for the Birdcage Quad, however it is surmised that the power gain is lower than that of a standard Quad as the spacing between the Birdcage elements is reduced by virtue of the elements being folded against each other. In any event, the Birdcage is a popular antenna in England and many overseas stations use it with great success. The Birdcage can be fed with a balun and coaxial line, or a balanced 70 ohm TV-type "ribbon" line and associated antenna tuner may be used.

MINIATURE QUADS

For those amateurs living in an apartment, a trailer or restricted quarters, the standard Quad may be too large for convenience. Attempts have been

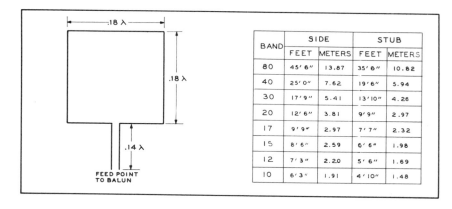

BAND	SIDE		STUB	
	FEET	METERS	FEET	METERS
80	45' 6"	13.87	35' 6"	10.82
40	25' 0"	7.62	19' 6"	5.94
30	17' 9"	5.41	13' 10"	4.28
20	12' 6"	3.81	9' 9"	2.97
17	9' 9"	2.97	7' 7"	2.32
15	8' 6"	2.59	6' 6"	1.98
12	7' 3"	2.20	5' 6"	1.69
10	6' 3"	1.91	4' 10"	1.48

Fig. 6 Mini-Quad loop, suitable for use on the low frequency bands, is made of smaller-than-usual Quad, with excess wire folded into a balanced transmission line. Total wire length is equal to that of a full size Quad element. Bandwidth of Mini-Quad is sufficient for 40 meter band and for 200 kHz portion of the 80 meter band. Stub may be adjusted to change resonant frequency.

made to "shrink" the Quad elements by the use of loading coils placed in the loops; the results have been inconclusive, and little specific comparative data has been found on the performance of the loaded Quad antenna. A coil-loaded mini-loop element was tried at W6SAI for a period of time to determine if it was competitive with a half-wave dipole mounted at the approximate mid-height of the element. After a period of tests, it was concluded that the loaded loop compared poorly with the dipole as far as signal strength reports were concerned and furthermore, the bandwidth of the loaded loop was quite restrictive compared to that of the reference dipole.

Further tests revealed that a mini-loop having about 75% the circumference of a standard quarter-wave Quad element could be stub loaded at the base, and would provide noticeable gain over the comparative dipole antenna. It was concluded that a Quad antenna designed about such an element would provide good power gain in a smaller-than-normal configuration. While a complete mini-Quad beam was never built, the dimensional information provided in Figure 6 may be of interest to an experimenter wishing to build a Quad that occupies a minimum of space.

Low Frequency Mini-Quads

The principle of stub loading may be advantageously applied to low frequency Quads, which otherwise become quite heroic in size when built for the 40 or 80 meter band. Using the data given in Figure 6, a Quad element for 40 meters is reduced to about 25 feet on a side, while an element for the 80 meter band is about 46 feet on a side — not particularly small to be sure but much more compact than a quarter-wavelength Quad which would run about 67 feet on a side!

Stub-loaded Quad elements may be adjusted with the aid of a dip-oscillator. The stub of the driven element is shorted at the feedpoint to complete the circuit and the stub length is varied to resonate the element at the design frequency. The length of the stub in the parasitic element is adjusted so as to tune the reflector about 5 percent lower in frequency (5 percent higher in frequency if the parasitic is a director) than the chosen design frequency.

The loading stub should be brought away from the element at right angles and may either drop beneath the loop or be brought back towards the supporting structure of the Quad.

As with any Quad, care must be taken if metal support arms are used to insure that the arms are not resonant near the operating frequency of the antenna. Long support arms should be broken at the midpoint with an insulator to prevent spurious resonance from occuring.

The low frequency Quad should be mounted with the center of the elements at least one-quarter wavelength above ground, or gain and front-to-back discrimination will suffer.

The 1-1/2 Wavelength Quad Loop

The regular Quad employs a closed element one wavelength in circumference. The X-Q array, on the other hand, uses an element two wavelengths in circumference. In addition to these two sizes, it is possible to employ a 1-1/2 wavelength loop in a Quad design, as illustrated in Figure 7. This *mini-X-Q* antenna configuration will have more gain than a standard equivalent Quad and somewhat less gain than the larger and more bulky X-Q array. A 15 meter mini-X-Q, for example, is about the same size as a 20 meter Quad of normal dimensions.

The "Monster Quad"

A few amateurs have experimented with five and six element Quad antennas on 14 MHz. This entails an array having a boom length of 40 to 65 feet with 10 to 13 foot spacing between the Quad loops. Such a *Monster Quad*, when checked out on a model antenna range, provided a power gain up to 11.5 decibels over a reference dipole, or 13.5 decibels over isotropic. This represents a power gain of about 14, which makes a kilowatt transmitter sound like a 14 kilowatt block-buster.

Normal element dimensions and spacings apply to an antenna of this size; the formidable problem is purely a mechanical one, that of assembling, erecting and keeping in the air (!) an array of such heroic proportions.

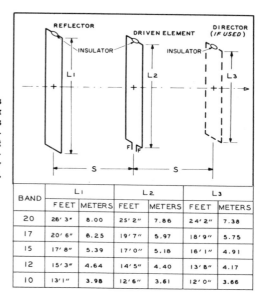

Fig. 7 Expanded Quad uses loops 1-1/2 wavelength on a side. The larger size provides greater power gain than standard Quad. Loops are broken at top for proper current distribution. Driven element has low impedance balanced feedpoint.

BAND	L₁		L₂		L₃	
	FEET	METERS	FEET	METERS	FEET	METERS
20	26' 3"	8.00	25' 2"	7.86	24' 2"	7.38
17	20' 6"	6.25	19' 7"	5.97	18' 9"	5.75
15	17' 8"	5.39	17' 0"	5.18	16' 1"	4.91
12	15' 3"	4.64	14' 5"	4.40	13' 8"	4.17
10	13' 1"	3.98	12' 6"	3.61	12' 0"	3.66

For a four element Quad, a 20 to 30 foot boom is often used (Figure 8). The boom may be made of two or three twelve foot sections of 2-1/2" o.d. x .065 wall, 6061-T6 (61-ST6) aluminum tubing. The sections are joined together by a 2-foot long section of tubing machined to slip-fit within the boom sections. The joints are pinned with 1/4-20 galvanized (or stainless steel) machine bolts and lock nuts. Various commercial Quad clamps and hardware are available to fit this size of boom tubing. The assembly should be strengthened by a top guy wire running between the boom ends and over a vertical support placed at the center of the boom.

Element dimensions are itemized on page 46. When assembed, the Quad may be placed atop a tall step ladder or other support and the elements checked for resonance with a dip-meter. The Tri-Gamma feed system discussed in Chapter VII is recommended for use with the Monster Quad.

The Quad arms may be made of fiberglas poles or aluminum tubing. If fiberglas is used, hollow poles should be filled with fiberglas plugs and epoxy cement to strengthen the arms at the point where they are held by the center clamp. Aluminum arms made of 1-1/8" and 1" aluminum tubing should be broken by an insulator at the mid-point of each arm to arrest spurious electrical resonances in the arms. The insulator may be made of a short hardwood or phenolic dowel machined to fit the inside diameter of the tubes.

The feedpoint impedance of a four-element Quad of this type may be adjusted to 50 ohms to provide a good match to a 50 ohm coaxial line.

Quad Versus Yagi — Is There A Difference?

A direct comparison between the effectiveness of a Quad and that of a Yagi is difficult to make and the results of on-the-air checks between the two types of antennas are often inaccurate and confusing. Laboratory field strength measurements of the power gain over a dipole of either type of beam antenna may lead to results that are controversial and open to various forms of interpretation. Unless such tests are run with extreme care on a well calibrated antenna range, the difference in power gain between the two antennas of approximately similar size will be lost in the inherent measurement error of the test set-up. Antenna tests repeatable to an accuracy of a decibel or better are difficult to manage even under the best of circumstances.

Extensive tests over the years imply that the Quad exhibits a power advantage over the Yagi (having an equal number of elements) of about 1.7 decibels. This advantage drops to about one decibel for large arrays, as shown in Figure 9. Thus a 2-element Quad is about equal to a 3-element Yagi, a 3-element Quad is equal to a 4-element Yagi, and so on.

OK — But Is It Worth It?

Is this advantage of 1.7 decibel or so in power gain that the Quad shows over an equivalent Yagi worth the effort of building the Quad, which some frustrated builders classify as a mechanical monster? After all, a Quad is a three dimensional object having length, width and height. The Yagi, on the other hand, is a two dimensional object having only length and width. Addition of the third dimension (height) to a building project immensely increases assembly and erection problems and also increases the wind resistance of the array. The Quad is a bulky, unruly, hard-to-handle assembly, usually of fragile construction. The Yagi antenna, on the other hand, is simple to assemble, rugged, and easily moved about on the ground and atop the tower.

The authors have used both the Yagi and the Quad over the years and have good friends using both types of beams with whom they have compared signals, both receiving and transmitting, under all types of radio conditions. In addition, they have had the opportunity of operating in various DX locations, listening to the signals emitted by Yagi and Quads — with the added advantage of knowing the antennas and operators of many of the stations worked.

Fig. 8 Tri-band four element Quad of W6CHE. Boom length is 30 feet. Element dimensions are given in Figure 7, page 46. The Tri-Gamma matching system (shown in Figure 1, page 76) is used. Typical SWR curves for this antenna are shown in Figure 4, page 94. An auxiliary brace hangs from the boom at the driven element to support the gamma matching sections and the various capacitors. Element arms are fibreglass poles mounted to aluminum "spiders". Boom is 2" diameter heavy wall aluminum tubing. The Quad exhibits good front-to-back ratio on 20, 15 and 10 meters and low SWR on all bands. Antenna is supported by 60 foot unguyed steel tower.

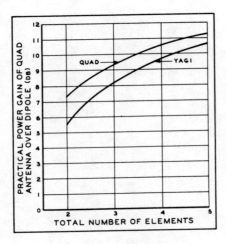

Fig. 9 A comparison between the Yagi and the Quad antenna. For an equivalent number of elements, the Quad exhibits a power advantage over the Yagi of about 1.7 decibel. Expressed in terms of boom length, the Quad may have about two-thirds the boom length of a Yagi for equal power gain.

THE MONSTER QUAD WINS!

The conclusion drawn from both objective and subjective tests over the years is that the Quad antenna has a definite advantage in terms of signal strength over the Yagi antenna, as suggested by Figure 9, Chapter X. the extra decibel or so *does make a difference over the long run.* Under conditions of difficult propagation, the Quad seems to outperform the Yagi, element for element, most of the time in a manner that is not readily explained by a mere comparison of antenna gain. Sometimes the Yagi seems to be better than the Quad, but the reverse seems to be true under more circumstances and over a longer span of observation. Generally speaking, the Quad antenna seems to "open the band" somewhat earlier than the Yagi, and "keeps the band open" a little longer than the Yagi. While a 5-element Quad may seem to exhibit only 3 decibels more power gain than a 3-element Yagi, the effectiveness of the Quad on the air, compared to the Yagi is truly formidable, and the results obtained with such a Quad often belie the apparently modest boost in power gain.

On the other hand, to take down a 3-element Yagi and go to a 3-element Quad, in the authors' opinion, could be a waste of time, as the small advantage gained may not be worth the money and effort spent in making the change. To go from a 3-element Yagi to a 4- or 5-element Quad, however, would truly open up a new world of DX performance.

THE W6SAI THEORY OF ANTENNA GAIN

Well, what is the secret of the Monster Quad? The W6SAI *Theory of Antenna Gain* seems to sum it all up: The DX-ability and overall effectiveness of any beam antenna is directly proportional to the time and money spent on the antenna, the difficulty of erecting it, and the overall size and weight of the array.

Surely, the Quad beam wins hands down on all of these points. What is true with automobiles evidently is also true of antennas — horsepower counts — and horsepower costs money!

The owner of the 3-element or 2-element Yagi, however, is not swept aside by the Monster Quad, no matter what the size and power gain of this impressive beam antenna. The ionosphere is no respector of antennas, and often the Yagi will beat out a Quad when impartial observers would guess otherwise. W6SAI, with his small 3-element beam, has no trouble in DX competition even though he may be bruised in a pile-up on occasion. That's what makes a horserace!

SWR Measurements

The various SWR curves for antennas described in this Handbook were measured at the end of a half-wavelength of coaxial line which was coupled to the antenna under test through a 1-to-1 balun. Care was taken to prevent coupling between the outer shield of the line and the field of the antenna.

SWR measurements can vary appreciably on a given antenna depending upon the accuracy of the SWR meter, the degree of unwanted coupling between the antenna and the outer shield of the line, and the proximity of nearby objects to the antenna. For best results the coax line should drop down directly beneath the antenna to ground level and the line connected to the antenna through an appropriate matching device. The transmission line should be coiled into a 5 turn inductor about a foot in diameter immediately below the antenna to help decouple the line from the antenna field.

Indicated SWR will vary with line length because practical SWR meters are not perfect and because isolation of the outer conductor of the line is not infinite. In the case of solid state equipment, experimentation with transmission line length may overcome loading problems but will not reduce the true value of SWR on the line.

OTHER BOOKS FOR RADIO AMATEURS, CB OPERATORS, SHORTWAVE LISTENERS, STUDENTS, & EXPERIMENTERS

BEAM ANTENNA HANDBOOK, by William I. Orr W6SAI and Stuart D. Cowan W2LX; 271 pages, 205 illustrations.

This popular new edition gives you: correct dimensions for 6, 10, 15, 20, and 40 meter beams; data on triband and compact beams; the truth about beam height; SWR curves for popular beams from 6 to 40 meters; and comparisons of T-match, Gamma match, and direct feed. Describes tests to confirm if your beam is working properly, tells how to save money by building your own beam and balun, and discusses test instruments and how to use them. A "must" for the serious DX'er!

THE RADIO AMATEUR ANTENNA HANDBOOK, by William I. Orr W6SAI and Stuart D. Cowan W2LX; 191 pages, 147 illustrations.

This clearly written, easy to understand handbook contains a wealth of information about amateur antennas, from beams to baluns, tuners, and towers. The exclusive "Truth Table" gives you the actual dB gain of 10 popular antenna types. Describes how to build multiband vertical and horizontal antennas, Quads, Delta Quads, Mini-Quads, a Monster Quad, DX "slopers", triband beams, and VHF Quagi and log periodic Yagi beams. Dimensions are given for all antennas in English and Metric units. Tells how antenna height and location affect results and describes efficient antennas for areas with poor ground conductivity. Covers radials, coaxial cable loss, "bargain" coax, baluns, SWR meters, wind loading, tower hazards, and the advantages and disadvantages of crank-up tilt-over towers.

SIMPLE, LOW-COST WIRE ANTENNAS FOR RADIO AMATEURS, by William I. Orr W6SAI and Stuart D. Cowan, W2LX; 192 pages, 100 illustrations.

Now-another great handbook joins the famous Cubical Quad and Beam Antenna Handbooks. Provides complete instructions for building tested wire antennas from 2 through 160 meters-horizontal, vertical, multiband traps, and beam antennas. Describes a 3-band Novice dipole with only one feedline; the "folded Marconi" antenna for 40, 80, or 160 meters; "invisible" antennas for difficult locations-hidden, disguised, and disappearing antennas (the Dick Tracy Special, the CIA Special). Covers antenna tuners and baluns, and gives clear explanations of radiation resistance, impedance, radials, ground systems, and lightning protection. This is a truly practical handbook.

THE TRUTH ABOUT CB ANTENNAS, by William I. Orr W6SAI and Stuart D. Cowan W2LX; 240 pages, 145 illustrations.

Contains everything the CB'er needs to know to buy or build, install, and adjust efficient CB antennas for strong, reliable signals. A unique "Truth Table" shows the dB gain from 10 of the most popular CB antennas. The antenna is the key to clear, reliable communication but most CB antennas do not work near peak efficiency. Now, for the first time, this handbook gives clear informative instructions on antenna adjustment, exposes false claims about inferior antennas, and helps you make your antenna work. With exclusive and complete coverage of the "Monster Quad" beam, the "King" of CB antennas.

INTERFERENCE HANDBOOK, by William R. Nelson WA6FQG; Editor: William I. Orr W6SAI; 253 pages, 152 illustrations.

This timely book covers every radio frequency interference (RFI) problem, with solutions based on years of practical experience. Covers amateur radio, CB radio, and power line problems with proven solutions. Contains case histories and lists valuable tips for stereo and TV owners to cure interference. Covers mobile, telephone, CATV, and computer problems as well.

ALL ABOUT VHF AMATEUR RADIO, by William I. Orr W6SAI; 172 pages, 107 illustrations.

Covers VHF propagation and DX, the VHF repeater and how it works for you, VHF moonbounce work, and amateur satellite communication. Discusses vertical and horizontal mobile antennas, tells almost everything about coaxial cables, and describes VHF beam antennas you can build yourself. Covers SWR measurements, VHF interference and how to suppress it, even care of the Ni-cad battery. This is a complete handbook of VHF radio for amateurs.

ALL ABOUT VERTICAL ANTENNAS, by William I. Orr W6SAI and Stuart D. Cowan W2LX; 192 pages, 95 illustrations.

Properly designed, built, and installed vertical antennas do a fine job in small places. This clear, well illustrated book covers the design, construction, installation, and operation of 52 vertical antennas: efficient Marconi antennas for 80 and 160 meters, multiband verticals, vertical loops, phased arrays, and shunt-fed towers. Also described are "radio" and electrical grounds, matching systems, tuners, loading coils, and TVI, plus the precautions necessary to protect yourself, your home, and your equipment from lightning damage . . . and much more! It's the most practical, authoritative vertical handbook published.

These popular handbooks save you time, trouble, and money in getting the most out of your equipment and your hobby. They condense years of study and successful experience into clear and interesting texts to help you obtain maximum results.